The Sovereign Entrepreneur

Cornell Studies in Political Economy
EDITED BY PETER J. KATZENSTEIN

The Sovereign Entrepreneur

OIL POLICIES IN ADVANCED AND LESS DEVELOPED CAPITALIST COUNTRIES

MERRIE GILBERT KLAPP

CORNELL UNIVERSITY PRESS

Ithaca and London

First published 1987 by Cornell University Press.

International Standard Book Number 0–8014–1997–2
Library of Congress Catalog Card Number 86–19909
Printed in the United States of America
Librarians: Library of Congress cataloging information
appears on the last page of the book.

The paper in this book is acid-free and meets the guidelines for
permanence and durability of the Committee on Production Guidelines
for Book Longevity of the Council on Library Resources.

To my parents

Contents

7

Preface

Why have governments in seventy-four countries become entrepreneurs in the oil industry rather than leave the business to the private sector? Why, moreover, have some governments established much larger public oil industries than others? The reasons behind the growth of public oil enterprise and variations in its scale, this book argues, are to be found in the changing role of the state; the limits to such enterprise stem from political opposition at home and abroad.

All governments have consistently advocated the expansion of state oil corporations to promote national welfare, legitimizing such expansion by means of a nationalistic and public rationale. Public enterprise is consistent with ministerial control for as long as sovereign rights, budgetary ties, and executive coalitions continue to bind the bureaucracy and the state oil company together. Governments vary in their achievements as entrepreneurs, however, and it is this variation that elucidates the very nature of the state itself.

Governments in less developed countries are able to expand their state oil companies more easily than can their counterparts in advanced industrialized countries. Crucial to the explanation of this difference, I argue, is the idea of pivotal bargaining power. Poor countries have by definition limited finances, which gives multinational corporations and international banks the pivotal power to limit the overseas operations of LDC state companies. Industrialized countries, on the other hand, are less susceptible to international pressure based on economic power; but there, domestic groups are much stronger than in less developed countries and can limit state enterprise in national operations through political opposition.

The pattern of oil industrialization in the capitalist world reflects this comparative logic. In this book I draw evidence primarily from case studies of Norway, Britain, Indonesia, and Malaysia, but I have checked it against the experience with oil of the United States, Japan, France, Italy, Mexico, Saudi Arabia, and Iran—countries that range from those with fully integrated state oil companies with worldwide operations to those with no state company at all. I conducted one hundred sixty interviews in 1978 with officials in national government ministries and international organizations, and with high-level representatives of industry associations, corporations, and trade unions. In order to extend the insights initially gained from these many informants, I have kept abreast of changes in the "oil patch" since 1979, by working with secondary materials and documents of various kinds.

This book makes three contributions to the theory of the state. First, it clarifies from a cross-national perspective the role of the state and the growth of state enterprise in oil. Second, it shows how ideas about the state do not have to assume an authoritarian state: conflicts within the state are consistent with statist theories. And third, the book extends the notion of state autonomy. State interests can remain autonomous, I demonstrate, even when the government is split by different particularistic interests or dependent upon external political groups.

The book thus goes beyond the existing literature on the state in several different ways. It systematically explains the growth of state control by relying primarily on political and bureaucratic rather than on economic variables. Drawing on Max Weber and other theorists of bureaucracy, the argument recasts the politics of management, budget, and property rights by focusing on the state in the role of entrepreneur. The comparative analysis of domestic bargaining also provides evidence of an autonomy for the state greater than that typically recognized by statist and dependency theorists. Finally, the focus on bureaucratic bargaining shows how product-cycle and international bargaining theories are played out "inside" the host state.

Why did governments create state-owned oil companies between 1968 and 1985? Chapter 1 sets the stage for political-economic struggles among state, multinational, and domestic companies. Chapter 2 uses evidence from four cases—Norway and Britain, Indonesia and Malaysia—to argue for a state-centered, or "statist," perspective that focuses on the nationalistic expansion of the state bureaucracy through oil. Chapter 3 presents the oil policy alternatives that were available to governments and discusses which policies the four countries actually chose. Chapter 4 then uses the statist perspective developed in Chapter 2 to explain the policy choices outlined in Chapter 3. Chapter 5 draws out the implications of the argument for state oil enterprise in seven other countries: the United States, France, Italy, Japan, Saudi Arabia, Iran, and Mexico. Chapter 6, finally, contrasts the argument in this book with those of major theorists of international bargaining, comparative domestic politics, and bureaucratic administration.

A few definitions will help clarify the text. The term "Majors" refers to seven international oil companies—Royal Dutch Shell, Esso, Mobil, Texaco, Standard Oil of California, British Petroleum, and Gulf. Also known as the Seven Sisters, these companies dominated the international oil market up to the mid-1970s. "Independents," by contrast, refers to the plethora of U.S., Canadian, French, Italian, and other European- and Asian-owned oil companies that began to produce internationally, particularly after 1960. These companies generally seek to feed their refinery operations with their own supplies rather than rely entirely on supplies purchased from the Majors. Finally, "state-owned oil companies" are oil companies in which a government holds the principal shares of stock and officially controls company investments and policies. The company must produce, refine, or distribute oil products for public or corporate consumption, and the revenues of the public enterprise must relate to the enterprise's costs even though societal objectives may be more important than profit maximization.

I express my gratitude to several people who helped bring this book to fruition. The book is a tribute to the insight, incisive comments, and perseverant encouragement of Peter Katzenstein. Its readability reflects the excellent editing of Roger Haydon at Cornell University Press. I am also extremely grateful to Peter Cowhey, Raymond Vernon, and two outside reviewers for perceptive readings and comments on the argument of the manuscript. On early drafts my colleagues Lawrence Susskind, Donald Schon, Bennett Harrison, Martin Rein, Lloyd Rodwin, Lisa Peattie, Karen Polenske, and Robert Fogelson at the Department of Urban Studies and Planning at MIT gave thoughtful comments and encouragement. For my intellectual training I remain profoundly indebted to Ernst Haas, and also to Melvin Webber and Michael Teitz. I also appreciate the research assistance of Phyllis Robinson, Mara Gelbloom, Michael Thomas, and Shiliang Tu at MIT and the excellent word-processing abilities of Karla Stryker. The initial research could not have been done without a ten-month international travel grant from the Institute for the Study of World Politics and financial assistance from the Marine Policy and Management Program at Woods Hole Oceanographic Institution and the Department of Urban Studies and Planning at MIT. Earlier versions of parts of the argument and case materials have appeared in articles in *International Organization*, the *Journal of Commonwealth and Comparative Politics*, and Jonathan Aronson and Peter Cowhey, eds., *Profit and the Pursuit of Energy* (Boulder, Colo.: Westview, 1983).

Most of all, I am eternally grateful to my parents for their love, encouragement, and tireless readings of drafts and to my husband, Sy Friedman, without whose love and hand this book would never have been completed, nor the universe minimally coded!

MERRIE GILBERT KLAPP

Cambridge, Massachusetts

The Sovereign Entrepreneur

CHAPTER ONE

The Struggle for Control of National Oil

In the 1980s we see a shift toward the privatization of state enterprise. Governments are selling off their public holdings to bolster national treasuries. But some observers have questioned the long-range wisdom of divesting profitable state enterprises. They have been joined by powerful heads of state enterprises, especially in oil, who have opposed advocates of privatization in finance ministries and international banks. Nevertheless, the cry for private ownership of industries, including oil, still resounds.

Is privatization merely an eddy, or will it prove to be the main current of future national decisions about economic policy? This book looks into evidence suggesting that the current still runs in the direction of state enterprise and that privatization, at least as far as oil is concerned, is no more than a diversion. The ownership of industry will depend, in the long run, not on ideological preferences but on struggles between great organizations—states, multinational corporations, and domestic companies.

In this chapter I outline the argument in brief. Then I set the stage for my account of the international and national play of oil power and wealth among state, multinational, and domestic organizations.

Governments in most of the capitalist world became oil entrepreneurs rather than leave the oil business to others. They did so, I shall argue, because governments had their own interests in oil enterprise which were only weakly constrained by societal groups. This argument is based on two main findings. First, governments pursued autonomous, institutional interests both in the public control and national ownership of oil and in public profits to supplement oil taxes. These entrepreneurial interests were consistent with those of other state agencies for as long as sovereign rights, budgetary ties, and supportive coalitions continued to bind together the bureaucracy and the state oil company. But these state interests were often independent of the business interests of societal groups. The international interests of multinational oil companies (both Major and Independents) and international banks might have been better served by the private production of oil; the domestic interests of national industrialists, labor unions, and other local groups, by stiffer taxation and regulation by government.

Second, government capacities for entrepreneurship differ from one country to another. Governments in less developed countries were more capable than those in advanced industrialized countries of expanding the power and scope of their state oil enterprises. Domestic groups in industrialized countries employed parliamentary and electoral threats to state power in order to constrict the growth of state enterprise, even in national oil operations. In contrast, international groups in less developed countries focused on profitable international operations to make way for state control nationally. But these groups also launched financial threats that strengthened the market barriers limiting state entry into international oil operations. In sum, state autonomy in oil and the relatively greater entrepreneurial achievements of governments in less developed countries are my two main points.

I focus on the state as actor. In most countries a coalition of leaders in oil or energy ministries, presidents or prime ministers, and parliaments or congresses advocated the expansion of state-owned enterprises to promote national welfare. They thus established a basis for consensus among bureaucratic, executive, representative, and entrepreneurial bodies of the state. The creation of a state oil enterprise institutionalized state interests in the development of the national oil industry, in public control, in acquiring oil profits to supplement oil taxes, and in exercising managerial discretion regarding oil-related goods and services. But to what extent were state interests consistent with state entrepreneurial policies?

The answer depended upon three conditions: first, the extension of sovereign property and ownership rights to the company; second, the fulfillment of bureaucratic as well as entrepreneurial interests in finance and management through new corporate endeavors; and third, the retention of an advocacy coalition within the state through bargaining. This is the statist part of my explanation of state oil enterprise.

A consistency between interests and policies was not in itself sufficient, however, for states to achieve their goals through state enterprise. States had to make bargaining concessions to repel opposition from international and domestic groups. Product-cycle constraints, limiting both domestic financial capacity and local expertise, strengthened the bargaining position of multinational oil companies and international banks. Such international groups could wield power over governments by threatening to withhold critical loans and by delaying important contract negotiations. They threatened state oil companies in less developed countries more than in advanced industrialized countries. By contrast, domestic political and industrial groups were in a better bargaining position to constrain the growth of state enterprise in advanced industrialized countries. Such groups could wield power by withholding marginal votes of support in parliaments split over oil policy decisions.

My statist perspective expands state-centric and domestic-bargaining approaches to the analysis of multinational corporate investment abroad. But it also differs in important ways from the integrated analyses of bargaining among state, multinational companies, and domestic groups to be found in statist and dependency approaches. State interests are analytically separable from domestic and international group interests, I find, and bargaining does not necessarily result in compromised or dependent policy outcomes that bend state actions toward societal will. Moreover, bargains between the state and international or domestic groups are not continuous, I find, but sporadic. Thus state autonomy is periodically, but not constantly, checked by bargaining threats. Such bargaining threats are most effectively used when state entrepreneurial investments depend upon parliamentary votes or financial resources. Domestic groups can most easily withhold votes; international groups can most easily withhold financial resources. (I call this "pivotal bargaining power.") State leaders have to chart a course between both sorts of groups. Internally, they must engage in the bargaining among bureaucratic, executive, representative, and entrepreneurial bodies of the state itself; externally, they get involved in bargaining between the state and multinational and domestic groups.

The overall pattern of oil industrialization in many capitalist countries reflects a comparative logic that I present in this book. The evidence is drawn primarily from case studies of Norway, Britain, Indonesia and Malaysia, but I have checked it against the experience of the United States, Japan, France, Italy, Mexico, Saudi Arabia, and Iran. From the particular combination of domestic and multinational constraints on the expansion of state oil companies, I shall show, we can predict the relative power and scope of national and international operations. We can also predict the pattern of concessions that limit state entrepreneurial expansion. State concessions to domestic groups in industrial countries lead to operational limits on the expansion of the state company's oil production and supporting services at the national level. I call this the *domestic concession* pattern. At the same time, product-cycle constraints that international groups impose on financing, contracting, and marketing limit the international operations of state enterprises from less developed countries. I call this the *multinational concession* pattern. Goverments in advanced industrialized countries are handicapped by a combination of domestic and multinational concessions. The state oil companies of less developed countries have more power and scope than those of advanced industrial countries, and so less developed countries experience primarily the multinational concession pattern. Because of opposition from powerful national bourgeoisies and populist groups, however, the combined pattern may also appear in some less developed countries such as Brazil and Argentina.

THE INTERNATIONAL STRUGGLE FOR CONTROL OF OIL

Before the international struggle for control of oil got under way, the Majors—Royal Dutch Shell, Esso, Mobil, Texaco, Standard Oil of California, British Petroleum, and Gulf, known as the Seven Sisters—dominated the production and trade of oil and gas supplies throughout the Western world. During the 1950s and 1960s independent oil refiners and producers in the United States and Western Europe began to erode the position of these giants in the Middle East. By the mid-1970s a titanic struggle had been staged over the control of national oil. It was not the Majors vying for global power; rather, it was new contestants challenging the Majors. These contestants were governments in the guise of oil entrepreneurs which had teamed up with Independents. This mixture of sovereign power and entrepreneurial, financial, and technical capacity could match the strength of the Majors, allowing the team of governments and Independents to win back control of national oil industries. National control was the battle cry of these sovereign entrepreneurs. The mixture of sovereign rights and corporate power was their muscle.

The seeds of nationalism in the control of oil resources were sown in the 1920s and 1930s. The Italian government was the first to try to unseat the Seven Sisters, forming a fully state-owned oil company in 1926. The Mexican government created a state oil company in 1934, nationalizing the entire oil industry in 1938— because Mexico consumed all of its oil domestically, the government did not need the cooperation of the Majors for access to global trading networks. Venezuela took a more gradual approach, acquiring concessions and favorable trade terms from foreign oil companies. The government did not nationalize the industry until 1976.[1] We can see here the beginnings of a general pattern in state participation in the industry: early nationalizers isolated domestic production, later ones retained the assistance of the Majors to participate in international trade.

A less extreme form of nationalism in oil appeared after World War II. Instead of nationalizing the entire oil industry, governments created state-owned oil companies to take a national share rather than to replace all foreign operations. Where private domestic oil companies already existed, they joined state-owned companies in taking part of the national share. The effect was to keep nationalization within the corporate boundaries of state enterprise and then to use national powers to gain a competitive edge within national markets. It was in this way that governments supplemented their sovereign power as governors with corporate power as entrepreneurs. This partial nationalization had a dual benefit: it gave governments more control over oil supplies and profits while preserving the taxable production and trading operations of foreign companies.

During the 1970s the trend to establish state-owned oil companies expanded to seventy-eight countries of the capitalist world.[2] The prevalence of the phenomenon indicated that most governments were coming to see state enterprise as a legitimate way to gain more economic, political, and operational control over the development of their national oil industries. The spread of state oil enterprise also led to a general restructuring of the global oil industry. Government entrepreneurs and Independents began to erode the dominance of the Majors in the global oil trade. By the 1980s the Majors controlled only about half of the petroleum supplies traded internationally, a dramatic change from the start of the 1970s when they had exercised almost exclusive control over international trade.

The period between 1970 and 1985 saw a titanic struggle among state-owned, multinational (Major and Independent), and private domestic oil companies for dominance in national and international markets. The three sorts of companies competed in national arenas for oil exploration, production, refining, and distribution, and internationally they also competed for trade and distribution.

The struggle was an organizational play for comparative advantage. Governments had an obvious advantage in their business deals: they could use sovereign *and* entrepreneurial capacities. They could both administer and bid for national oil production, tanker contracts, and bilateral oil deals with other governments. Multinational corporations, by contrast, had their own advantages, principally in substantial capital and expertise and existing trade networks that governments lacked. In some cases—Royal Dutch Shell, for instance—these resources had been developed over eighty years of involvement in global production and trade. Experience and established networks continued to give the Majors a competitive edge in national and in global markets. The Independents also had the advantage of substantial capital and expertise, but they did not possess established trade networks. They generally were private domestic oil companies from the United States, Japan, France, Canada, Italy, and other advanced Western countries. In most cases their interest in producing oil in other countries was to gain access to additional supplies for their refining and distribution operations at home. Often the Independents accepted less favorable deals offered by host governments in less developed countries in order to offset the international trade advantages that the Majors could offer.

Finally, private domestic oil companies in advanced industrialized countries had their own advantage: they constituted the backbone of national economies and the political blood of governing coalitions. Private corporate groups could withhold tax payments or foreign exchange earnings from their governments. They could also threaten to sabotage government policies by mobilizing blocks of votes at the polls or in parliament. Governments could not do without tax income and foreign exchange to balance international payments, nor could they survive without the political support of important domestic industrial groups.

This comparative advantage of domestic companies was particularly influential within the national markets that governments administered for oil leases and subcontracting work. It had little effect in international markets, where domestic companies were pitted against multinational companies in competition over contract terms for price and supply.

This struggle for market shares of national oil is important because through it the organization and control of the oil industry were recast in the 1980s. During the struggle, oil supplies, oil profits, and industrial operations were divided up, inside national boundaries and among international networks. Two examples are illustrative. First, governments cooperated in the Organization of Petroleum Exporting Countries (OPEC) to increase prices and impose production ceilings; cooperation was critical, securing a major market share for OPEC by creating global scarcity of supply. Second, the prevalence of state-owned oil companies in non-OPEC countries led to greater reliance on agreements among producer and consumer governments than on deals between private corporate importers and exporters. These forms of cooperation led to a complete pattern of global production and trade involving intergovernmental, government-multinational, MNC-MNC, government-domestic company, and multinational-domestic company deals. Such changes seemed to promise new business networks, indeed a completely new oil industry.

By the mid-1980s oil was no longer a scarce commodity. How has the struggle for oil played itself out? Governments have successfully used state-owned oil companies to gain major shares of national production, transportation, refining, and marketing; state companies have thus eroded the relative share of the Majors. Governments have also tightened up their taxation policies to prevent multinational corporations from reinvesting profits elsewhere. Finally, governments have used their administrative control over oil leasing to prevent private domestic oil companies from gaining a significant part of the national share of oil operations. In these ways governments have carved out for themselves a major entrepreneurial role within national oil industries. But they have not achieved this new role without problems inside the bureaucracy. Often government leaders have been at cross-purposes with leaders of state oil companies about whether oil revenues should be reinvested or distributed for national economic purposes.

But the Majors have not lost out entirely. They have succeeded in shifting tax burdens on their profits to countries with lower tax rates. They have also reinvested earnings to benefit their operations overseas. For a time during the early 1980s some Majors acquired other oil companies (for instance, Standard Oil of California acquired Gulf) as a way to expand their global control of oil industry operations without having to negotiate new leases with governments directly. When dealing with governments, however, the Majors continued to drive tough bargains. In exchange for conceding national shares to state companies, the Majors often demanded control of oil exports, which enabled them to retain international control of oil supplies and overseas distribution.

Majors and Independents did not act as a unified group in the struggle for control of oil; they competed fiercely with one another. While the Majors were trying to retain their national production operations, the Independents were actively striking deals with the state oil companies of host governments in order to gain production opportunities for themselves.

Private domestic companies were the least successful participants in the oil struggle. They possessed neither the unique international characteristics of the multinationals nor the unique sovereign characteristics of state-owned oil companies. What financial and political influence they did hold over government policy, however, earned them a minor share in national production and transport operations. Domestic companies were able to use their financial contributions (foreign exchange and tax revenues) and their political support (their control over government coalitions) to gain some preferential treatment in government allocations of oil leases and oil-related subcontracts. A few private domestic groups, such as fishermen, were sometimes able to persuade governments to create policies that freed them from physical interference by the oil industry in offshore zones.

But only those private domestic groups which wielded critical influence over governing coalitions always got their way on policy choices. Such influence generally involved parliamentary or electoral votes or financial resources without which governing coalitions could neither remain in power nor continue to pursue their objectives in the oil industry.

Although these international and domestic conditions suggest similarities across countries, the unique political economic history of each country produced distinctive variations on the state-multinational-domestic struggle. In each country the contest was staged against a distinctive backdrop of government control, international finance, and indigenous national entrepreneurship.

COMPARATIVE POLITICAL ECONOMIES OF OIL ENTERPRISE

How did differences in national political economies alter the contest between multinational, state, and private domestic companies for dominance within national and global oil industries? If nationalistic sentiments in Indonesia and Malaysia, and much later in Norway and Britain, had led to national unity in oil, then state-owned oil companies might have joined with private domestic oil companies in order to erode the market share of the multinationals. Alternatively, if the ideological battle between conservative and state-interventionist groups was really the key factor in oil industrialization in Norway and Britain, then state-owned oil companies might have opposed growth in the market shares of both private multinational and domestic oil companies. If it was actually the cohesion of international capital groups which led to cooperation between foreign oil investors and governing elites in Indonesia and Malaysia, and later in Norway and Britain, then state-owned oil companies might have worked with multinational corporations to prevent private domestic groups from gaining shares in newly developing oil markets. If Indonesia's and Malaysia's development of oil at an early stage in the product cycle led to antagonisms with multinational corporations, then state-owned oil companies in Norway and Britain may have cooperated better with the multinationals because they waited until the 1970s, when even marketing barriers to entry were being eroded.

Norwegian Political Economy

At the turn of the twentieth century Norway was not among the leading industrial competitors for foreign investment. It joined the ranks of modern industrial societies only after World War I. But it remained poorer than most European countries, with a per capita income and a level of manufacturing per capita below those of even the poorest countries of eastern and southern Europe. Between the wars Norwegians relied on agriculture, commercial trade, manufacturing, and shipping for their livelihoods.[3]

Natural resources, such as oil, and indigenous domestic entrepreneurship signaled a potential for rapid national economic growth. Like most European countries, Norway regained the relative losses it had suffered between the two world wars during the growth sprint of the 1960s and 1970s. In this growth phase the country was ruled by social democratic governments committed to managing demand by simultaneously achieving full employment, price stability, and trade equilibrium. Norway's postwar growth culminated in a speculative boom in 1971–73 which coincided with the emergence of the Norwegian oil industry. Oil attracted shipping, chemical, and other indigenous industrialists, but those sectors which were not able to reap oil benefits (and particularly manufacturing) suffered from slower growth, rising unemployment, and inflation.

24

In this postwar period the state was cast to play a strong role in the Norwegian economy. With only one short intermission, in 1961–64, Labor party coalitions governed Norway from 1945 until 1981, when they were unseated by a coalition led by the Conservatives. The Labor government goals of full employment and high economic growth without sharp price rises or balance-of-payments problems could be achieved only with public-sector involvement in investment, consumption, and trade. But the government was unable effectively to regulate private lending institutions and private reinvestment of profits, and from this inability stemmed a relative lack of success (in 1952–61, for instance).[4]

Despite setbacks, government leadership strongly influenced private domestic investment after the war. It provided tax incentives for private-sector investment and stimulated industrial development in less developed northern areas of the country. The government achieved economic goals most effectively through state-owned banks; some credit restrictions were also applied through private credit institutions subject to state credit agreements or credit law. But the size and structure of the public sector and its influence on private business were not much greater than in other West European countries.

Unlike other European countries, Norway did not depend heavily on foreign financing until the mid-1970s, when its oil investments began to necessitate substantial borrowing overseas. Up to 1970, for instance, only one-fourth of the Norwegian fleet (fourth-largest in the world) was owned by foreigners.[5] Only hydroelectric power around the turn of the century and the aluminum industry during the 1950s had relied heavily on foreign financing. But Norway was extremely dependent upon international trade: after 1950 annual exports comprised about 40 percent of gross national product. According to Stale Seierstad, if the relation between import and export prices (that is, the terms of trade) had fallen 10 percent during the early 1970s, it would have forced Norway to reduce imports by an amount corresponding to 4 percent of net national product to keep its international balance of trade unchanged.[6]

Norway also veered strongly away from European norms in the matter of national sentiment. Norwegians believed that industrial growth should be owned by Norwegians and managed so as to retain a "balance" with other cultural values. Most industry, except for hydroelectricity and aluminum, was Norwegian-owned up to the mid-1960s, but thereafter the influx of foreign oil companies caused many Norwegians to worry that their socialist, environmental, and conservationist values might not survive. Opposition parties such as the Agrarian and Green coalitions grew stronger as oil development speeded up during the 1970s. National sentiment favored oil development, but not at the price of unbalanced growth in other sectors of the economy or of the displacement of people from jobs or housing. The fight for these principles against the temptations of oil wealth spurred Norwegian nationalistic sentiment against foreign oil enterprises after about 1970.

The unavoidable economic instability introduced by oil exacted a national political price. Depressed oil prices, especially in 1980–82, caused disequilibrium in the external balance of payments and in the government's internal budgeting. Government efforts to respond by reducing expenditures met opposition from those Norwegian industrial groups likely to suffer. It was during this period that the Labor party lost control of the government to the Conservatives. The loss of the 1981 election, according to Øystein Noreng, was "directly linked to the perceived failure of liberal energy policies" and of individual leaders and their political party. Government leaders have since tried to insulate the Norwegian economy from the harmful fluctuations of the global oil market in a desperate attempt to recreate political and economic stability within the country.[7]

These conditions within the Norwegian political economy fostered a strongly nationalistic stance on foreign oil investment from the late 1960s through the 1980s. The state's moderation in extending investment credit set the tone for public oil entrepreneurship to offset foreign oil investment during the 1970s. Then economic upsets were created as oil revenues and foreign oil employees poured into the country, turning Conservative, Agrarian, and environmentalist groups against the foreign oil companies. Finally, the modest role of international capital in the national economy grew to major proportions as foreign oil companies made extensive investments. In Norway, in sum, business and political conditions became polarized; in Britain, meanwhile, they stagnated. In both countries, however, the result by 1974 was strongly nationalistic in tone.

British Political Economy

British business, like its Norwegian counterpart, tolerated moderate state intervention in the economy, but it relied more heavily on international capital. Foreign capital did little to ameliorate the economic plight of the British people, however. So, as in Norway, nationalist sentiment regarding oil and jobs grew.

Although most wartime controls had been removed by the mid-1950s, the government continued to play a significant role in the British economy. By the mid-1960s more than half of the gross national product was being absorbed by all levels of government (if we include subsidies and grants) and by publicly owned corporations. Public corporations operated coal mining, electricity generation, railways, inland waterways, and atomic energy. The iron and steel industry, civil air transportation, and road services were also publicly owned. And agriculture was heavily subsidized.[8]

In most cases public ownership offset the negative effects of constant inflationary pressures, severe balance-of-payments deficits, and threats to the exchange value of sterling. What resulted, from 1956 to 1963, was a "stop-go" cycle in which government action stifled spurts of relatively rapid economic expansion. As a result the economy experienced low average rates of growth, about 2 percent per year, across the period.

Both Conservative and Labour governments were to blame. The Conservatives' monetary policy from 1951 to 1964 was no more successful than Labour's structural reform and moderate growth policy from 1964 to 1970. Labour was determined to speed up the modernization of industry, stimulate exports, improve the productivity of the labor force, and create a statutory prices and incomes policy. A major part of that structural overhaul was led by the government-financed Industrial Reorganization Corporation, which attempted to rationalize the organization of British industry by promoting acquisitions and mergers. But by 1967 Britain was in "fundamental disequilibrium." The country could not achieve a reasonable rate of growth and balance of payments at the existing parity values for sterling. Britain's share of manufactured exports had fallen progressively, while imports had risen persistently.

This crisis necessitated large-scale foreign financing to keep the economy afloat. The government was forced to devalue, and the International Monetary Fund and foreign central banks lent heavily to alleviate the balance-of-payments deficit. The government instituted tight fiscal and monetary policies that, helped by a boost in world trade, brought a surplus of foreign exchange to the Treasury in 1969. Helped by an upturn in world trade, this improvement in the country's trade balance was bought at the price of economic growth and high unemployment.

When the Conservatives came back to power in 1970, the government tried to hold off wage demands in the public sector. Inflation and labor unrest increased, and dock, local council, power station, and mining workers scored record wage deals. But in 1973 sharp increases in global oil prices quadrupled the cost of British oil imports and broke the back of the Conservatives' rapid growth policy. By 1974 the Labour party was back in power. Labourites abandoned the incomes policy and attempted to forge a new "social contract" with the country's powerful labor unions. They used substantial increases in taxation to offset rapid wage increases. But despite a rise in exports, the continued devaluations of sterling forced the government to borrow from overseas banks, further increase taxes, restrain government expenditures, and reinstitute an incomes policy.

It was during this period that the Labour government decided to intervene directly in British oil production. Public corporate power and the influence of the labor unions fostered intervention by the Labour government. British businesses such as shipping shuddered at this new element of state control. But with the foreign debt growing as international capital poured in to offset balance-of-payments deficits, there seemed to be little choice. The government would earn its own capital by generating oil profits.

General conditions—unemployment, slow growth, deficits, and devaluation of sterling—also fostered a strongly nationalistic sentiment. The British people, and national labor unions in particular, felt that British oil should be produced by the British, for the British. As in Norway, the presence of an interventionist Labour government and national unity behind the notion of British control of industry set the conditions for government involvement in oil. In Britain, however, unlike in Norway, international capital had played a major role in bailing out the Treasury and in financing key industries, dissuading the government from more than a minimal entrepreneurial presence in oil.

Indonesian Political Economy

The postwar Indonesian economy was relatively free of upsets caused by domestic industrialists and labor unions. There was little in the way of private national capital invested in domestic industries; most investment was foreign—American, Japanese, or Dutch. Many domestic industries, such as fishing, were still artisanal in organization. What national capacity did exist in such industrial sectors as shipping was state-owned. Moreover, the government, believing that labor organizations were linked to communist movements, strongly discouraged trade unionism. The Communist party was abolished when the pro-Western Suharto government took over in 1966 from Sukarno's socialistic government.

But the absence of strong, indigenous industrialists and trade unions did not mean that the Indonesian government was free to go its own way. The government was constrained by a strong nationalism. Nationalism was not, as in Norway and Britain, a sentiment rekindled by the presence of foreign multinational companies in oil development during the 1970s. Rather, Indonesia had experienced the presence of foreign oil companies and Dutch colonialists since the turn of the century, and even after independence from the Dutch in 1945 the Indonesians could not afford to rid themselves of foreign investment in oil. Sukarno prohibited foreign investment in all other industries. And while Suharto embraced the foreign oil companies, he also retained state enterprise investment to keep national interests from being overwhelmed by foreign interests. Nationalism was a strong thread throughout the fabric of Indonesian society, politics, and economics.

Nationalism was not the only political factor constant across both the Sukarno and the Suharto regimes. Sukarno relied heavily on military elites to fill his cabinets. Suharto, although he attempted to reduce the military presence and increase the involvement of economists and other technocrats, still filled key posts with army personnel. Suharto claimed that his close cooperation with the army was necessary to rout remaining elements of the Communist party.[9]

The Indonesian government has consistently retained strong, centralized control over the Indonesian economy. From 1945 until 1971 military elites ruled the country; they determined the allocation of investment funds to various industrial sectors and social programs in different regions of the country. The first general elections since 1955 were held in 1971, and as expected, the government's official party won, further strengthening Suharto's control. Suharto created a new parliament, which became dominated by four factions: the government party, the armed forces, Muslims, and socialists. He also introduced economists into key positions within the Ministry of Finance, BAPPENAS (the regional planning department), the Ministry of Mines, and other government agencies.[10] Through 1974 the Suharto regime enjoyed a period of political and economic stability.

Incidents in 1974 broke the tranquillity. Student riots stirred up criticism of the favoritism that the government showed toward foreign business interests and its own small group of governing elites. The government's strong links to American and Japanese business through the economists in the Finance Ministry drew particular criticism. Suharto responded to student unrest and subsequent concern among military groups by returning to the highly nationalistic approach of Sukarno. He modified his stance toward Western countries and capitalism in general, adopting the idea that a policy of nonalignment was the best way to ward off a communist takeover.[11]

The financial crisis of the state oil company, Pertamina, added to Suharto's problems. Suharto did not find out that the company had defaulted on its repayment of short-term debts to foreign banks until after the company was near bankruptcy. Government economists pressured Suharto to take direct control of Pertamina. Internal government struggle over the issue eroded the government's political position both nationally and internationally.[12]

By 1976 the political situation was so bad that the government decided to stage elections. Suharto insisted that opposition parties reorganize into two new parties, to favor a victory for the government's official Golkar party. Four Muslim parties merged to form the United Development party (PPP), and the three nationalist and two Christian parties combined to form the Indonesian Democratic party. These mergers created political parties that lacked the ideological coherence needed to contest the official platform. Golkar won, and the following year Suharto was elected for another five-year term.[13]

Indonesia has been politically stable since these elections. Golkar again won parliamentary elections in 1982, by a decisive majority, and a landslide victory in twenty-six provinces (only one held out for the PPP because of local Muslim strength).[14] In 1983 Suharto was reelected for another five-year term.

Despite this period of strong nationalism and strong control by the central government, international capital played an increasingly large part in the economic development of the country. At the start of Suharto's regime the economy was on the verge of collapse. Annual growth in gross domestic product averaged 2 percent between 1960 and 1966. Investment had declined, budget deficits loomed, and the productive capacity of the industrial and export sectors was minimal. The annual inflation rate was already 640 percent and increasing at a rate of 20 to 30 percent per month. Suharto immediately abolished many of Sukarno's prohibitions on foreign capital investment in Indonesia, opening the way for a "new era" of economic growth financed by foreigners.[15]

The three national five-year plans between 1969 and 1984 aimed to rehabilitate the economy and hold down soaring inflation. Replita I (1969–74) relied heavily upon foreign investment to reactivate private-sector activities in such areas as agriculture and infrastructure. Much of the $2 billion expended was foreign direct investment from the West and Japan. The plan succeeded in attaining a growth rate of 9.4 percent per year, in large part because of this rapid influx of foreign capital, revenues from growth in oil production, and the strict political grip of the military.[16]

The second plan, Replita II (1974–79), started amid growing national dissatisfaction with the government's encouragement of foreign investment. The plan was modified to stress national concerns about job creation, housing, income distribution, regional development, and education. Thanks to windfall oil revenues that accrued during the OPEC price increases of the 1970s, the plan met most of its goals.[17]

Replita III (1979–84) recognized that a much lower rate of

growth in the economy should be expected, about 5 percent, owing to serious declines in the price of oil. The plan continued to stress job opportunities, import substitution, strengthening of the export base, and growth in agriculture-based and manufacturing industry.[18]

Whereas Labour governments in Norway and Britain had been strongly influenced by domestic private business or labor union interests, the military government in Indonesia retained much more centralized control. Nationalism was the unifying political message but the economic necessity was heavy international capital investment, needed to sustain a rate of growth in gross domestic product at between 4 and 9 percent during 1969–84.[19] The ability of the government to undermine opposition groups by reorganizing political parties insured the continued rule of Suharto and his government party.

Malaysian Political Economy

The story in Malaysia resembled that in Indonesia. Nationalism, a strongly centralized state leadership, and a critical role for international capital from the start characterized Malaysian political and economic development.

Of the four countries examined here, Malaysia has the shortest history as an independent nation. British colonial rule began in the late eighteenth century. During World War II the country was occupied by the Japanese and then was subject to a major Communist insurgency during the 1950s. But in 1957 independence was declared from the British and a constitutional monarchy established. A federal government and parliament were also created to govern the new Federation of Malaya. In 1963 the federation was expanded to include the former British crown colony of Sarawak and protectorate of North Borneo (now Sabah) and renamed Malaysia.

From the start the federal government was strongly centralized under the leadership of the prime minister. Although ill-advised to act without the support of the king and the nine sultans of the Malaysian states, the prime minister retained executive power over the cabinet and was the leader of the political party that won most seats in the parliamentary elections. Except between 1969 and 1971, when parliamentary democracy was suspended, the Alliance party and later the National Front has been the dominant political force in Malaysia. The Alliance party ruled until 1969 and then merged with six opposition parties in 1973 to form the National Front (Barisan Nasional). Between 1974 and 1985 prime ministers Razak, Onn, and Moham-ad ruled with majorities of more than 70 percent in Parliament. Such overwhelming parliamentary support enabled successive prime ministers to lead Malaysia during the period of state involvement in oil with unquestioned authority.

The nationalism that had stirred political unrest during the 1960s was aimed primarily at the dominance of foreign capital in Malaysia. In 1971 native ("Bumiputra") Malaysians, other Malaysians of Chinese and Indian descent, and foreign companies controlled respectively 4 percent, 34 percent, and 62 percent of corporate equity in the country. In order to quell domestic opposition to this inequity and strengthen its own political position, the government adopted the New Economic Policy in 1971. The policy aimed to channel future growth in the economy to Bumiputras without transferring any existing equity among groups. The goal was to achieve by 1990 a corporate equity distribution of respectively 30 percent, 40 percent, and 30 percent for Bumiputras, other Malaysians, and foreigners. Another aim was to increase the percentage of Bumiputras employed generally, and in managerial and technical positions in particular.[20]

This policy was a double-edged sword for foreign investment in Malaysian industries. It deterred investment by restricting direct foreign investment and the management and employment policies of foreign companies. But the country attracted foreign investment by promising a stable political environment in which investors could be guaranteed a return on their investment without fear of unrest or government upheaval. The New Economic Policy also worried Chinese and Chinese-Malaysian investors. Although the policy nominally protected their share of investments, the legislation's political aim was to give Bumiputras more economic control within the domestic economy. It would increase the government's share of public enterprise in the name of disadvantaged Bumiputras and perhaps build up some private Bumiputra capacity. But it might also threaten existing Chinese and Chinese-Malaysian investments in shipping, fishing, and other sectors.

Foreign investment accounted for much of Malaysia's prosperity. Early industrial development had included large-scale production of rubber, timber, and tin. But by the 1970s petroleum and palm oil, and by the 1980s natural gas, had become important industrial sectors. During the 1970s the domestic economy experienced real growth averaging 7 percent per annum, and inflation was kept to 5–6 percent, low by world standards. But the country was highly dependent upon exports: the ratio of exports to gross national product was over 50 percent. Rubber and tin accounted for two-thirds of export earnings in 1961; oil, palm oil, and manufactured exports reduced that share to only half in 1970.[21]

But by 1980 Malaysia's balance-of-payments deficit had begun to grow rapidly, spurred by global recession, falling commodity prices, and rising import prices that reflected world inflation. The government's increasing public expenditure also contributed to the deficit. Public investment grew more than 16 percent per year from 1975 into the 1980s. Petroleum exports and commodity export taxes had allowed the Malaysian government to expand its oil and gas investments during the 1970s, but the subsequent fall in oil revenues and in tax receipts produced a rise in deficits. Deficits rose to 21 percent of gross national product in 1982 but declined marginally in the following two years.[22]

The countervailing forces of Malaysian nationalism, foreign investment, and Chinese-Malaysian and Bumiputra participation strengthened the government's roles as a manager of economic imbalances among ethnic groups and as an entrepreneur on behalf of native Malaysians. The strong central government played a role similar to that of its counterpart in Indonesia. But Malaysia differed from Norway and Britain, where private domestic industrial groups were politically and financially strong. In Malaysia, although the investment of Chinese-Malaysians was important to the domestic economy, these groups had little political influence.

Norway, Britain, Indonesia, and Malaysia were chosen as cases because they exhibit important similarities in oil industrialization as well as differences distinctive of advanced industrialized and less developed countries. All four countries were involved not only in oil industrialization but also in developing offshore as well as onshore oil resources. New, offshore production markets and the potential for extensions of sovereignty opened up opportunities for state enterprise which avoided nationalizations of private operations. State enterprise can thus be studied as a competitive policy choice in a market environment rather than as merely the authoritative use of force by a government.

The choice of Norway and Britain, on the one hand, and Indonesia and Malaysia, on the other, was made to accentuate differences between advanced industrialized and less developed countries. Norway and Britain are both modern industrialized countries with long-standing representative democracies. Indonesia and Malaysia, in contrast, are both newly industrializing countries, their statehood a result of postwar independence from Dutch and British colonial regimes. Techno-bureaucratic-military governments hold centralized power in both countries, although formal representative bodies do exist.

Similarities within each pair of countries also contributed to the choice, to make even clearer the variation between chosen oil policies. Both Norway and Britain have relatively recently developed oil in the North Sea. Their political and economic positions within the global economy have also been similar. And they have comparably strong industrial sectors (shipping and fishing) that compete with oil for industrial space and economic importance in their domestic economies. Both Indonesia and Malaysia, by contrast, saw their oil resources developed late in the nineteenth and early in the twentieth century. They also have similar international positions as newly industrializing countries in Southeast Asia. Their industrial sectors competitive with oil are either government-owned (shipping) or artisanal (fishing); both shipping and fishing hold comparable status in the two domestic economies.

These similarities and differences across the four cases are substantial. What explanation of the state's role in oil industrialization and entrepreneurship can interpret them in a coherent theoretical light? It is to this question that we now turn.

CHAPTER TWO

A Statist Perspective on Public Enterprise in Petroleum Resource Management

Why did the Norwegian, British, Indonesian, and Malaysian governments, along with seventy-four others around the world, become entrepreneurs between 1960 and 1985 in order to manage the exploitation of petroleum resources?

Radical changes occurred in the international oil market across that quarter-century—unstable prices, uncertain supplies, and proliferating new oil companies. They suggest that conditions, although conducive to government intervention, did not compel governments to create public oil enterprises. Previously, high capital costs in production and refining, high fixed costs in operations, and short-term elasticities of supply but inelastic demand had created barriers to market entry which secured for seven major oil companies over 90 percent of world oil trade. This dominance combined with periodic applications of U.S. and British diplomatic pressure to keep prices and supplies stable.[1]

After 1960 structural changes began to occur in the international oil market. OPEC countries started to increase prices artificially and suppress production in order to retain their relative share of overall non-Communist oil supplies. That share increased from under one-half in 1973 to two-thirds by 1980.[2] The initial price rise in 1971, a quadrupling of prices in 1973, and a doubling of prices in 1978–79 signaled to oil-importing countries such as Norway, Britain, and Malaysia that it was possible to raise prices while intentionally keeping production low, creating an artificial condition of

36

oil scarcity. OPEC countries were also able to reduce supplies available on the market by placing embargoes on oil exports or by nationalizing oil production.

Precarious price and supply conditions fostered a proliferation of new oil companies anxious to capture windfall profits by selling oil at near-OPEC prices. The Majors had controlled 90 percent of oil trade in 1973, through equity and buyback agreements. These agreements tied oil products into distribution networks or third-party sales arrangements. By 1980 this market control had decreased to only 55 percent. The 10 percent of direct sales on the spot market or in processing deals controlled by governments or independent oil companies in 1973 had increased to 45 percent of the international oil trade by 1980.[3]

These market conditions would not by themselves have led to public entrepreneurship. Government leaders had to have some motivation to enlarge the state's role from that of sovereign to include that of entrepreneur. In this chapter I focus on the role of the state to explain the emergence and survival of state enterprises in capitalist countries. I expand and modify a statist perspective.

A STATIST PERSPECTIVE

A "statist" perspective has emerged from the efforts of scholars over the past twenty years to analyze states as principal forms of political organization guiding international political economic events. These scholars generally agree that the interests and organizational coherence of states are often fragmented by the influences of societal groups outside their boundaries.[4] Nevertheless, states have demonstrated that they have interests of their own and can act as cohesive units to make decisions. The general state-as-actor model has been summed up by Richard Ashley as

an entity capable of having certain objectives or interests and of deciding among and deploying alternative means in their service. Thus, for purposes of theory, the state must be treated as an unproblematic unity: an entity whose existence, boundaries, identifying structures, constituencies, legitimations, interests, and capacities to make self-regarding decisions can be treated as given, independent of transnational class and human interests, and undisputed (except, perhaps, by other states).[5]

It is within this general framework that the emergence and expansion of public entrepreneurship within national and international oil economies can best be explained. Governments could not have created and sustained public enterprises in the face of opposition from society, had governments not been able to separate their oil interests from purely societal interests. This shift enabled them to redefine their managerial roles to include corporate ownership and profit making.

I accept this general statist perspective because of evidence presented in this book, but I modify and expand it to deal with three ambiguities not sufficiently explained in the statist literature. The first ambiguity concerns the definition of the state's interest: Can the state's interest be different from that of society? The second ambiguity is whether the state can be autonomous and effective as an entrepreneurial actor: Are bureaucracies, freed from the influences of societal groups, able to pursue a unified set of interests and to implement those interests consistently in successive state policies? The third ambiguity is whether a statist perspective can be combined with an international bargaining perspective: Is there a comparative statist logic of state-multinational-domestic bargaining?

THE STATE'S INTEREST

The first ambiguity in the statist perspective lies in the definition of the state's interest and concerns whether the state can have an interest different from society's. Generally scholars have not been willing to dissociate state interests from societal interests, and so the argument that states act partly in their own bureaucratic interests rather than always for societal interests fundamentally alters the conventional state-as-actor perspective. According to Robert Gilpin, the state—composed of individuals who hold authority—has interests of its own and can be autonomous. But it effectively represents broader societal interests by constituting a "coalition of coalitions whose objectives and interests result from the powers and bargaining among the several coalitions composing the larger society and political elites."[6] Gilpin argues that any separation of government from societal interests is a reification of the state. This perspective may, however, be obfuscating an important theoretical distinction that needs to be made in order to interpret evidence about public entrepreneurship.

In this book I use "the state" to refer to the institutions of government. These institutions are bureaucratic agencies and executive offices, including such state-owned bodies as public companies and banks, in addition to such representative branches as parliament. State and society together comprise the nation-state.[7]

Such a definition of the state, as both governor and entrepreneur, is consistent with Max Weber's notion that the state has an institutional authority and legitimacy separate from and superior to the society and the individuals that it governs. Weber claimed that the state held binding authority over its own public officials, its citizens, and activities within its national territory. The state could use this compulsory jurisdiction, its own continuity as an organization, and its legitimate force to make decisions as a sovereign. Weber thus argued that the state had an impersonal authority (he termed it "rational-legal") that was independent of individuals but extended to them if they were officials. His notion of bureaucratic authority suggests that the state's decision to become an oil entrepreneur might be an autonomous act of leadership by bureaucrats representing the state as an institution.[8]

By "autonomy" I refer to the organizational loyalties of a coalition of officials acting to fulfill the institutional goals of the state.[9] Such a coalition may include the president or prime minister, heads of state banks or state enterprises, and members of government departments such as ministers of energy, industry, trade, commerce, and the environment. A coalition of these officials may place the broader interests of the government (including parliament) or the bureaucracy above the special interests of particular individuals or ministries. These broader interests involve industrial growth, social welfare, and the financial and political needs of the bureaucracy as a manager of the national economy.

Weber also argued that bureaucracy is the most efficient form of social organization for capitalist production.[10] The argument seems to suggest that the entrepreneurial state may represent an advanced management stage in the development of the oil industry. Bureaucratic management would be perceived by public officials as a "precise, stable, and reliable" way to organize a domestic oil industry.

The cases of Norway, Britain, Indonesia, and Malaysia, as I will argue in the following chapters, show that state interests in oil entrepreneurship favored bureaucratic ownership and management of national oil. But ministers and parliaments interpreted these interests as independent of the benefits that accrued to particular domestic or foreign companies.

In Norway the minister of industry was able to convince Conservative and Liberal parties in Parliament that Norway needed a direct supply of domestic oil. Domestic supplies would offset the country's dependence on world supplies. The ministry also convinced Parliament that Norway needed direct government control and ownership so that national oil and gas resources would not be exploited by foreign companies. Øystein Noreng argues that in Norway the macroeconomic interests of the state have generally taken precedence over the profit interests of private oil business. State enterprise was justified as the best way to serve the public's need for secure energy supplies and the nation's need for greater national control over national resources and industrial development.[11]

The rationale was similar in Britain, but the British National Oil Corporation (BNOC) was also advertised as a solution to the country's financial troubles. By 1970 Britain was suffering from its highest inflation rate, highest unemployment, and largest balance-of-trade deficit since World War II.[12] Ministers of energy and finance argued that state enterprise could reduce the trade deficit by producing domestic oil for the British economy. It could also assure economic growth, employment, and oil-related business contracts for British labor and industry.[13]

Government bureaucrats used similar rationales in both Indonesia and Malaysia. They justified public oil companies in terms of reduced energy dependence, increased national control within the oil industry, and benefits for local labor and businesses. But government ministers argued most persuasively for the need to expand national industrial capacity. National industrial development was important in Norway and Britain, but particularly in Indonesia and Malaysia it was strongly tied to nationalism. The historical experience of Dutch and British colonialism had planted a strong belief in Indonesia and Malaysia that independent national institutions were essential. Only independent institutions could protect the growth of national industry from a foreign exploitation reminiscent of colonial days. Ministers in both countries saw public oil companies as instruments to build an oil industry that would be nationally controlled rather than foreign-dominated. In addition, public oil companies could earn foreign exchange—exchange that would reduce national dependence on international capital markets to finance national industrial development.[14]

As this brief summary suggests, ministers in all four countries justified the use of public enterprise by ascribing attributes of the "national interest" to the unnegotiable rights of the state acting in the "public interest." The "national interest" has traditionally referred to an interest of the nation-state which depended upon bargaining and persuasion, one in which the state was not necessarily in a superior decision-making position. Had the four countries accepted the negotiability of the national interest, they would have left state oil enterprise open to a debate pitting public enterprise against private alternatives advocated by foreign governments (such as the United States) and the Majors.

But in the traditional "public interest" the state acts in the interests of the public as a superior authority. It chooses among alternatives for society according to its social contract as sovereign.[15] For instance, the government decides to whom to allocate oil leases. By assuming this superior position on administration, entrepreneurial states could justify preferential leasing to state companies in the name of the public's oil interest. This government capability legitimized public oil entrepreneurship through the exercise of an unchallengeable public authority composed of the leasing and preferential bidding rights of ministries and state enterprises. Such public authority affected development of national oil industrial capacity, oil supplies for national energy and security purposes, and the use of oil as a way to shift economies toward public control (all reasons consistent with a mercantilist perspective). This combination of national and public interest is what Bruce Andrews refers to as an autonomy of the state that becomes commensurate with the national interest:

> An autonomy, a self-regarding, self-sufficient motivation, takes shape within the sphere of the state itself. Government becomes an extra-social category, a virtually self-referring unit, tethered only rhetorically . . . to a society's general welfare. . . . What the central government is consistently interested in achieves the status of the national interest.[16]

States in all four countries thus developed new rationales for public entrepreneurship. Ministerial interests in oil income and management control were legitimized by a combination of national interests in oil development and public interests in economic and natural-resource control. How bureaucrats interpret these national and public interests over time defines the state's "interest" as I use the term. What differentiates the state's interest from the bureaucracy's interest is its national industrial rather than civil service or administrative focus. What differentiates the state's interest from the government's interest is that the former may remain unchanged despite changes in leadership.

The central hypothesis of this book, formally stated, is that the degree to which bureaucrats emphasize autonomous state interests affects the extent that they choose public enterprise as their oil policy. The hypothesized effect is contingent upon the degree to which domestic or international private-sector groups in society oppose state decision making (bureaucratic or parliamentary) by withholding finances or votes that are critical for governing coalitions to remain in power. Such opposition can shift state interests back to a societal orientation. The more domestic the societal opposition, the less extensive will public enterprise be. The more that societal opposition comes from international groups, the more extensive public enterprise will be in national oil operations, but the less extensive in international operations. This distinction between national and international operations has implications both for the presence of public oil companies and for the extent of their vertical integration.

CAN THE STATE BE AUTONOMOUS AND CONSISTENT?

The second ambiguity in the statist perspective addresses the relationship between state autonomy and consistency. Stephen Krasner and Peter Katzenstein have both categorized states as "strong" or "weak" depending upon their ability to "resist" or "neutralize" opposition from societal groups. Katzenstein shows that states in small West European and Scandinavian countries can neither act autonomously nor have political interests independent of those of the private sector, because the challenges of domestic political or producer groups "neutralize" but do not undermine state efforts.[17] This conceptual distinction of state from society is important to our understanding of state autonomy even if state interests are neutralized.

Despite the compromise of state autonomy, Katzenstein demonstrates, state strength is actually increased or reduced by the interactions of state and society in democratic corporatist regimes. Strong ideologies of social partnership build cooperation among the state, private business, and labor groups. The centralized and concentrated peak associations and informal, voluntary bargaining also build consensus among interest groups, the state bureaucracy, and political parties.[18] In Austria the dominant role of political parties in these cooperative political arrangements has reduced the power of the relatively strong, centralized state to that of a hesitant entrepreneur and leader of nationalized industry and nationalized banks. But Katzenstein finds that the same kind of consensus politics led to a strengthening of the weak, decentralized state in Switzerland. The need both to arbitrate between and to unify bureaucracy, unions, and internationally oriented business forced the state into a leadership role.[19] Although they ultimately rule out state autonomy, corporatist state-society relations thus produce state strength.

The examination of coalition-building processes emphasizes the emergence and maintenance of state strength as a social process. But the perspective obstructs a more autonomous definition of state interests, because it portrays a state and society meshed together by constant bargaining.

Other observers have been more willing to grant that particular state agencies may exhibit autonomy in certain policy areas (but not that this autonomy characterizes the state as a whole). Krasner finds the United States to be a weak state that achieved autonomy in its foreign policy for raw materials investment. In this issue-area geopolitical concerns over foreign military intervention and ideological concepts of U.S. hegemony clashed with powerful corporate interests. Policy making was therefore isolated within the state in a small group including president and secretary of state and their respective bureaucracies, the White House and the State Department. But these state agencies were not entirely insulated from societal pressures, because of the relative influence of Congress on various parts of the federal executive. The U.S. case, according to Krasner, demonstrates the capacity of weak states to exhibit isolated, sectorally specific characteristics of autonomy.[20]

In more recent work Krasner makes a related argument at the international level by examining rule-changing autonomy rather than merely organizational autonomy. He argues that weak states in less developed countries can improve their positions in the world economy. They can do so if they succeed in either introducing new authoritative rules or increasing their gains from existing market rules in a particular sector such as oil.[21]

But these analyses retain vestiges of earlier, society-centered views. Krasner does not interpret state strength or autonomy as an organizational characteristic of the state-as-institution. Moreover, the idea that individual sectoral policies can be inconsistent with overall state strength underestimates the impact of intersectoral politics. Both views are necessary to explain state entrepreneurial achievements in oil in the face of societal opposition to state enterprise and sectoral opposition from fishing and shipping.

Another approach to state autonomy is to focus on the discretionary and discriminatory capabilities of the state as an industrial manager. John Zysman argues that states must be able to exercise "discretion" in allocating industrial finance credit to influence private-sector strategies as a rival or partner would. This capacity, along with the ability to "discriminate" among firms contingent upon their actions, is essential for the state to pursue continuous and systematic industrial objectives.[22] But Zysman does not highlight the state's ability to expand its management discretion through a state corporate structure that would facilitate an even more autonomous role for government.

State entrepreneurship offers the potential for both state strength and autonomy. In his early work Stuart Holland examined the conditions under which governments could use state enterprise as a strategic policy instrument of national planning. Referring primarily to the Italian Institute for Industrial Reconstruction, Holland showed that a government could use state companies to insure expansion, capital deepening, increased output, and capital-intensive processes in sectors where low-growth market conditions would normally require state loans to stimulate expansionary private investment. He also argued that government ownership of fewer than half of a company's shares would enable it to control the investment, location, and pricing of the company as well as of competing firms in an entire industrial sector. The government could use this latter, oligopolistic strategy to manage growth according to capacity and demand even in sectors in which the state company produced less than half of sectoral output.[23] This was a compelling argument for state-led industrial growth through state enterprise. But by focusing on economic factors, Holland neglected the bureaucratic consensus and political support necessary for such an expansion of state enterprise within the economy.

Other authors have more recently claimed that nationalistic rather than simply planning motives lead governments to replace foreign investment and technology with state enterprise. John Freeman suggests that state entrepreneurship is a "nationalistic response to the problem of achieving advanced forms of industrialization in the context of extensive dependence on foreign sources of capital and technology." The reliance on state enterprise, he argues, changes the institutional structure of the state by transforming the composition and interests of bureaucratic agencies.[24] National bourgeoisie, technocratic elite, and international groups become incorporated into the state itself, while public-sector administrators and managers take on new interests and build new political bases.

The key to a statist explanation of state enterprise, I argue, is whether a strong state can pursue consistent interests over time. Can strong states, despite changes in ideologies, partisan opposition, or governing coalitions, retain sufficient bureaucratic autonomy and unity to sustain entrepreneurial oil policies over the long run? Three conditions determine the consistency of state interests and action. The first is the extent to which the state extends its sovereign property rights to its own companies. The second is the extent to which the state formulates new corporate interests and forms of management discretion that continue to serve the interests of key ministries. The third condition is the degree to which bargaining within the state sustains the unity of state coalitions supporting state enterprise.

But objectives must be not just consistent over time but also effective. I investigate "effective state action" here in terms of the market share and the extent of vertical integration of oil investments that governments achieve acting as entrepreneurs.[25] This is a measure of state power, not of efficiency. The greater the share and extent of vertical investment, the greater will be the ability of the state to manage the national economy. Entrepreneurial investment can represent autonomous state interests, common interests with private-sector groups, state interests with concessions extracted by politically influential groups, or some combination of these elements. Such state investment represents only one aspect of the state's managerial relationship with civil society, of course, and Chapter 3 looks at others, among them taxation, regulation, and leasing.

We can approach the state's consistency as an autonomous actor by analyzing the relationship between state interests and state entrepreneurial action. State interests are under the stewardship of officials in the state bureaucracy; state action here refers to the investment goals of state-owned oil enterprise. Consistency can thus be examined in the chains of command between bureaucratic officials and heads of state enterprises.

Two hypotheses are plausible. One stems from recent work by Raymond Vernon about the link between ministers in government and the managers of state enterprises.[26] His focus on ministers is more limited than the broader notion of the state advanced here, which also includes the rest of the bureaucracy—presidents or prime ministers, heads of state oil companies and state banks—and members of parliament. However, for the moment let us consider only the link between ministers and state corporate managers in order to explore Vernon's hypothesis.

Vernon appears to assume that ministers have diverse interests because of their various responsibilities for finance, energy, trade, or the environment. If such diverse interests reflect societal concerns about these areas, the state that such ministers serve could be considered weak. The greater the diversity of ministerial cross-pressures on the state enterprise, the less are state company managers able to achieve their multiple and often conflicting goals. My own hypothesis, drawn from the four cases examined in this book, assumes that ministers have unified and consistent interests in forming state oil enterprises. Their collective concern is for what is best for the nation. The greater the ability of ministers to sustain this unity, the more consistent will be the set of goals that corporate managers achieve over time. If Vernon's hypothesis is plausible, then it is difficult to argue that states are strong, because they cannot pursue autonomous and consistent sets of goals through oil entrepreneurship. If my hypothesis is plausible, however, then the autonomy and consistency of strong states will be a useful notion in understanding oil matters.

Referring to the French case, Vernon argues that dual tutelage reduces the ability of the bureaucracy to act as an arbitrator of national wants. When both a ministry for financial issues and a ministry for technical matters were placed in charge of the French state enterprise for electric power, the action increased the chances for countervailing or even conflictual interpretations of the national interest. Mixed commands may imbue state enterprises with multiple, diverse, and often conflicting objectives, with several ministries able to grant rewards or punishment. Prevailing uncertainty in the business environment only increases the likelihood that ministers will change their directives to state enterprises over time.

Once they are in operation, state enterprises have two sources of new interests. First, as in the French case, state enterprises may be instruments of the nation as a whole, not subordinate to the interests of the particular bureaucracy in power. If national interests change, so do those of the state enterprise, and they change irrespective of bureaucratic needs. Second, as in the Japanese case, state enterprises may change their interests by a "succession of marginal adjustments" that "accommodate the rights and interests of dissenting groups" including private business, the bureaucracy, and political parties. In sum, according to Vernon, the idea of a unified, coherent set of ministerial directives to state enterprises is remote.

> In operational terms, then, the agreements that operate between managers and ministers usually consist of an eclectic mix—a mix that commingles long-term gains with short-term objectives, that changes frequently in content, and that is rarely tested for its internal consistency. When efforts are made to put such agreements in formal terms, as France has sought to do, they cannot fail to reflect those eclectic characteristics. The idea of a rational set of goals, responding to some coherent concepts of optimality and serving as a feasible measure of command and control, remains remote.[27]

The second hypothesis, by contrast, takes the perspective that ministers are in consensus on the unified set of objectives which state oil enterprises are to achieve. The greater the unanimity among ministers, the more consistent will be the objectives achieved by the state oil enterprise. Consistency indicates a shift from the diverse substantive concerns of ministers to a unified focus on the bureaucratic interests of the state itself as a manager of the national economy.

The ability of ministers in Norway, Britain, Indonesia, and Malaysia to pursue consistent interests through state oil enterprise depended, I argue, upon three conditions internal to bureaucracies. Whether these conditions were met over time determined whether the link between ministers and state oil company managers was eclectic and contradictory or unified and consistent. These conditions are the ability of ministers 1) unilaterally to create a new government company, backed by parliamentary legislation, and grant it sovereign rights to exploit national oil and gas resources; 2) to expand the administrative discretion of ministries by taking on a new corporate discretion in matters of profit making, ownership, contracting, and operational control; and 3) to define a charter for the public oil company which would give it autonomy within the state yet sustained backing from other agencies including parliament. Governments in all four countries were remarkably similar in meeting these conditions.

Sovereignty was the essential "given" of state power which allowed ministers, backed by heads of state and members of parliament, unilaterally to create government oil companies and grant them exclusive rights to national oil and gas resources. Sovereignty, however, is a characteristic of the state itself; it cannot be parceled out to various ministries. An act of the state, formally unifying its ministries, is required in order for sovereignty to be used. This is true whether sovereignty is vested in the nation (France), the crown (Britain), or a constitution (the United States).[28] Three features of sovereignty are unique and give those who act in the name of the state a power not available to private actors. The state has the right to use military force to take control within its territory; it has the right to determine independently the limits of state power without the control of another state or government agency; and it has the right to be independent or "self-sufficing" because it controls a particular population and geographical area.[29] Sovereign power is supreme in both the international and the domestic sphere as long as no other states, on their own or on behalf of corporate actors, challenge that power by the use of force. Ministers and state corporate heads thus have supreme decision-making power as long as they act formally in the name of the state. In that name they can also increase state wealth and territory more than in any other way—except through war, imperial conquest, or the formation of nation-states.

Parliaments and ministers can therefore extend property rights consistent with sovereignty to the state, a diversion from the state's typical use of sovereignty (which is to grant property rights to others). Public choice theorists such as Douglas North argue that the state specifies property rights or "rules of the game"—laws, regulations, contracts—only for private companies, not for itself. The state attempts to maximize its rent from private exploitation (in the form of tax revenues) while private companies try to minimize transactions costs that are imposed by state enforcement of private property rights. The result is an inefficient allocation of property rights to private groups, because the state can maximize its oil tax returns only if transactions costs—which contribute to efficiency—are kept low.[30]

But the state's use of sovereignty to grant property rights to state companies proves to be a deft acquisition of oil revenues and control, more effective than the use of taxation. By becoming an oil company with sovereign property claims, the state moved from merely maximizing taxes to maximizing its own oil profits, directly owning operations, and participating in oil contracts and leases. It also gained opportunities for a monopoly access to resources and a vertical integration of national oil industry operations within its own corporate structure.

The Norwegian Ministry of Industry proposed to the Norwegian Parliament that Statoil "take advantage of the rights acquired under the contracts relating to government participation as a basis for engaging in transport, refining and marketing . . . [and] thereby play a major part in realizing the Government's policy of establishing an integrated Norwegian petroleum community."[31] Similarly, the British Parliament in the Petroleum and Submarine Pipelines Act of 1975 charged the British National Oil Corporation with performing the operations of a fully integrated oil company, including exploration, production, refining, distribution, and petrochemical production throughout the world.[32]

Governments in the two less developed countries were even more forceful in asserting sovereignty through state oil company rights. The Indonesian government gave Pertamina a parliamentary directive, in Law no. 44, that "mining undertakings of mineral oil and gas are exclusively carried out by State Enterprises" (Art. 3). "Mining undertaking" was defined in Article 4 to include all operations of an integrated oil company including exploration, production, refining, transportation, and marketing.[33] The Malaysian government modeled its state oil company after Pertamina. Parliament passed the Petroleum Development Act of 1974 in which it referred specifically to a management share for the national oil company in the production, transportation, and refining of Malaysian oil and gas.[34]

States thus relied on the hierarchical or vertically integrated structure of state-owned firms. To the extent that they did so, they could extend their sovereignty from authority over territory to authority over profit making, industrial ownership, and discriminatory contracting. Extended sovereignty was the first condition for a consistency of objectives for government ministries and state oil enterprises.

The second condition was that the state company take on "new" interests to expand management by ministries. Traditionally, ministers have tried to maximize their administrative discretion by maximizing their finances. William Niskanen argues that heads of ministries—or bureaus—seek to maximize their budgets, which are often subject to grants approved by representative branches such as parliament. In this exchange of ministry services for budgetary grants the shrewd minister can gain leverage to bargain for a higher budget by exploiting differences between the service demands of the executive and the legislature.[35]

A state oil company gives ministers a new opportunity for corporate discretion in management. Oil profits can increase ministerial budgets, while corporate ownership, vertical integration, and subcontracting discretion can expand the state's management control within the oil industry. So although state oil companies in all four countries initially assumed the interests of the governing coalitions that had established them, the implementation of corporate oil policies led to new state interests. I borrow this argument from Douglas Bennett and Kenneth Sharpe, who stress that even though state interests may be historically shaped, the actual pursuit of certain public policies can lead to new state interests.[36]

The consistency of state interests from case to case depends upon how state oil companies used their new profit-making and management discretion. Such discretion included the ability of state companies to allocate their own oil profits, production savings from economies of scale in operations, marketing and tax savings from vertically integrated multinational operations and transfer pricing, and investment capital borrowed from foreign banks. As long as state companies maximized ministerial or military budgets and control with revenues from these sources of income, interests within the state remained consistent, and conflict among ministries, the military, and state oil companies was therefore minimal. Such was the initial relationship between state oil companies and ministries in the four countries. In Norway, for instance, Statoil remained closely tied to the directives of the Petroleum Directorate; in Malaysia, Petronas remained close to the prime minister's Economic Planning Unit in charge of oil affairs. In Indonesia the army more explicitly traded its political support for the state oil company, Pertamina, against side-payments to the military.

But as state oil companies began to act more in their own interests as corporations, and less in the budgetary interests of ministries, state interests became less consistent. This intrastate competition for finances and management control often produced direct conflict.

Conflicts occurred between state oil companies and ministries of finance, planning, and energy, and the military, over oil payments to the national treasury, investments, subcontracting objectives, and foreign borrowing. In Indonesia, for instance, Permina, predecessor to Pertamina, directed foreign oil companies during the 1960s to pay oil taxes and royalties directly to it rather than to the Indonesian Treasury. The state oil company used this capital and its own oil profits to finance new oil-related investments as well as investments in schools, hospitals, hotels, roads, and shipping. This infuriated the ministries of Finance and of Planning (BAPPENAS).[37]

In sum, the consistency between ministerial and state oil company interests depended upon whether the oil company used its new corporate capacities to contribute to (but not substitute for) the budgetary and management capacities of state ministries. Once ministries and state oil companies became competitors for oil revenues and control, conflicts emerged within the state. These conflicts could be resolved only by rebuilding internal coalitions.[38]

This brings us to the third condition necessary for the state to pursue consistent goals through state oil enterprise. Oil policies of state oil companies need to remain fairly constant, despite changes in state leadership or feuds inside and realignments of coalitions within the state. This predictability of oil policy depends upon the ability of ministers to nurture supportive coalitions within the state. It also depends upon the ability of those coalitions to remain strong despite ideological shifts of government and unstable international oil markets.

52

In all four countries ministers of industry and energy, prime ministers, or presidents were able to generate unified support from other ministers within the state for wholly state-owned oil companies. But, surprisingly, the initiators of state enterprise did not immediately seek to exercise their influence. Instead, to the new companies they handed charters or lists of government directives which gave the companies substantial autonomy from the influence of all ministries, including those that had initially supported their formation.[39]

Once established, state oil companies managed to sustain oil policies consistent with their original charters, despite changes in governing coalitions and ministerial or state company leadership. In Britain, for example, despite the electoral change in 1979 to a Conservative government that was ideologically opposed to state ownership in any industry, the British National Oil Corporation was never entirely privatized. Instead, the Conservatives restructured the company to create Britoil, still a state oil company with minority rather than majority state shareholding. The state company was simply too useful to the Treasury as a generator of long-term profits—whether from sales of oil or from sales of stock. It was also a convenient way to acquire foreign loans that did not appear to add to an already bloated public-sector debt load.

But policy consistency also depended upon the state and ministerial interests being collectively served. State institutional interests were served as long as the presence of a state oil company appeased nationalistic sentiments and oil profits continued to bolster national income. State oil companies also served state interests if they met national employment and industry requirements and as long as oil industry development remained centrally managed through the role of the state company. On the other hand, particular ministerial interests continued to be served as long as state oil companies paid treasuries a share of their profits and directed oil-related loans to aid government policies for other parts of the economy. The state oil company could appease the Energy Department in Britain, for example, if it made at least minor attempts to hire British industry and labor for its North Sea operations.

It was on the continuing service of national and ministerial interests that coalitions were built within the state. Oil companies could sustain or even realign existing coalitions of government leaders, ministers, and parliamentarians in favor of the basic oil policies being pursued by the state oil company. In some cases this service involved new forms of financial payoffs to treasuries or new forms of operational and administrative control for energy and finance ministries.

So despite threats and some privatization by Conservative governments in Britain and Norway, and despite a reorganization of the state oil company in Indonesia, state oil companies in all four countries retained fairly consistent oil policies. They avoided, or even to a moderate degree sustained, extreme fragmentation and conflict within the state without substantial impact upon their oil policies.

To sum up, three conditions had to be met if state oil companies were to pursue consistent rather than eclectic state institutional interests. First, sovereign property and ownership rights had to be extended to state companies. Second, both the bureaucratic and the entrepreneurial interests of the state had to be fulfilled by the new corporate financial and management gains of the state company. Finally, advocacy coalitions had to be sustained or rebuilt within the state in order for the enterprise to survive recurring onslaughts from finance and planning ministries.

The evidence from the four countries lends credibility both to the hypothesis extrapolated from Vernon's work and to my own. State autonomy in oil represents at times diverse and at times consistent interests linking ministers and state company managers. The eventual conflicts that occurred between ministerial directives and the interests pursued by state oil enterprises support Vernon. Ministers and state corporate heads were united in initial support for state oil companies. That unity began to erode, however, as various ministers, especially of finance, tried to emphasize their particular agency interests over the national interests that had originally appeared to justify state enterprise. But diverse commands did not lead to confused goals for state enterprises; rather, heads of state oil companies stuck firmly to their initial charters and to their newly acquired interests as competitive corporations. The fact that state oil enterprises continued to pursue consistent interests, despite the growth of opposition from ministers, supports my own hypothesis. State enterprises can pursue autonomous, consistent sets of oil policy objectives as long as a strong coalition of ministers and the executive continues to support their efforts.

Some portion of the ministries and state enterprises of a strong state may advocate and pursue consistent, autonomous interests of the state in oil. But other ministries and the parliament of the strong state may reflect eclectic or conflicting goals, other bureaucratic purposes (for example, minimizing the public deficit), or societal interests that oppose state enterprise in oil. The more the state pursues interests and entrepreneurial goals that are inconsistent, the less autonomous it becomes.

Bargaining between State and Society

The third ambiguity is how a statist perspective relates to an international bargaining perspective. Can a comparative logic be inferred from conditions under which domestic and multinational coalitions constrain the growth of state oil enterprises?

The cases in this book suggest that patterns of state entrepreneurial investment can indeed be systematically analyzed. These patterns are based not just on the strength of states themselves but also on the presence of strong or weak domestic and multinational coalitions. Decentralized, representative democratic governments in Norway and Britain were more vulnerable to the influence of domestic societal groups than were centralized governments in Indonesia and Malaysia. In the latter two countries, however, international groups influenced government interests when public enterprise concerned international oil industry operations.

Two different bargaining principles enable societal coalitions to extract concessions from the state. First, societal groups use pivotal votes and finances to curtail the growth of state enterprises. "Pivotal power" is a principle of bargaining power which grows out of the conventional recognition that swing votes are influential in balance-of-power elections. Second, the coalition can extract concessions regarding national or international oil operations from the state (depending upon whether it is is composed of domestic or multinational actors). As this book demonstrates, state-society bargaining in different countries can be systematically compared in use of "pivotal power" and in domestic or multinational character of societal coalitions.

Bargaining Shifts

The assumption that the state and society have different interests is the basis for a bargain between them. (In "society" I here include both multinational and domestic private companies—a break from conventional usage.) To the extent that branches of the state such as parliaments represent societal interests, bargaining among bureaucratic and representative state bodies reflects the state-society split. Such internal state bargaining emerges especially as oil agencies gain autonomy from other state bodies.

The bargains that were reached in the 1960s and early 1970s were asymmetrical. States were weak or lacked sufficient capital or expertise in oil to pursue entrepreneurial oil interests of their own. They had to make bargains with multinational oil companies, because in general private domestic industrialists also lacked sufficient oil expertise. The basis for the bargain was an exchange of "access" to oil reserves for finances, expertise, industrial capacity, or votes. States controlled access to oil on national territory. The multinationals possessed the finances, expertise, and oil industrial capacity. Domestic groups could influence parliamentary and electoral votes.[40]

In all four countries, governments had sovereign claims to oil and gas resources both onshore and offshore, through bilateral treaties such as the Norwegian-U.K. delimitation treaty of 1965 and the Indonesian-Malaysian agreement of 1969.[41] Governments could restrict access to the resources they controlled by giving preferences to certain companies or by limiting the number of leases put up for bid during licensing rounds. Agencies could also prohibit oil trade, freeze funds, use force, and veto the extension of capital for loans and investment.[42]

But despite their control of resources and territory, states could not afford the investment capital, oil drilling and production technologies, and expertise to exploit those resources. By the early 1970s Britain was suffering its worst balance-of-payments crisis since World War II and could hardly divert financial resources to oil development. Indonesia and Malaysia also were already heavily dependent upon loans from the Japanese, U.S., and European governments, independent Chinese and American banks, and such international banking and loan organizations as the International Monetary Fund and the World Bank. Only Norway was in a strong financial position at the beginning of its oil industrialization. It was not until the mid-1970s that Oslo became indebted to foreign banks to finance its ambitious plans for oil development.[43]

Without substantial imports of financial and technical capacity, therefore, most governments could not have started national oil industry investments. Initially, this condition of affairs was the basis for an asymmetrical interdependence between the state and foreign oil companies.[44] Over the 1970s, however, these asymmetrical bargaining relations shifted, making states less dependent upon the multinationals. Governments formed state oil companies and used those companies to earn oil profits and seek foreign loans to buy drilling technologies and expertise.

Bureaucracies in the four countries jumped at the chance to use public oil companies to secure loans from foreign banks, which enabled bureaucrats to acquire international capital tied neither to foreign oil company investments nor to loans from foreign government. Norway's state oil company, Statoil, was initially financed by loans from the Norwegian government and the Bank of Norway. Britain's BNOC was fully financed by the British Treasury. By 1978, however, both oil companies had received direct loans from foreign banks: Statoil about $400 million from a group of foreign banks, and BNOC $825 million from American banks in 1977. This external borrowing was justified in the Norwegian case as a way to import capital without increasing Norway's substantial accrued public debt. That debt was expected to rise to about $20 billion by 1978. In Indonesia, meanwhile, American banks such as Citibank and Chase Manhattan invested billions of dollars in Pertamina's oil operations. Malaysia's Petronas was financed by the Malaysian government, which received direct loans from Chinese and Japanese investors.[45]

Bureaucrats also used oil enterprise to gain access to the technology and expertise of those foreign oil companies willing to enter business partnerships with the state. The Majors were reluctant, except in Malaysia, but Independents formed consortia with public oil companies in order to gain access to oil production in Norway, Britain, and Indonesia.

Several factors strengthened the bargaining power of the bureaucracies, as Theodore Moran's work suggests.[46] Oil technology involved high fixed investment and high fixed cost. Once operations were producing, firms could no longer use the threat of withdrawal as a bargaining tactic. Furthermore, the promise of oil wealth made bureaucracies and nationalistic groups ambitious in their bargaining demands. Finally, the hectic competition among Independents and Majors for access to oil resources enabled host governments to discriminate among foreign firms.

Barriers to entry in international oil market operations continued to decline over the 1970s, facilitating new public oil company investments. The degree of vertical integration in the industry, the concentration in the world market, and the oligopolistic power of world purchasers also began to decline. As a result, the relative importance of marketing barriers decreased.[47]

Supply shortages in the international oil market also contributed to the obsolescing bargain that had begun to emerge in the 1960s and was dominant by the mid-1970s. According to Vernon, "as shortages appeared in various raw materials, multinationals lost the bargaining power that their marketing capabilities normally afforded." The bargaining position of governments and state-owned oil companies relative to that of the multinationals was thus strengthened.[48]

The shift from oil shortages in the 1970s to glut in the early 1980s should have eliminated at least the international conditions favoring state entrepreneurship. Falling prices should have reduced the availability of capital with which public oil companies could overcome barriers to entry. Nevertheless, public oil companies survived through the mid-1980s in Norway, Britain, Indonesia, and Malaysia. They formed partnerships with foreign oil companies (through joint ventures and long-term service contracts) to overcome increased competition for oil sales in a volatile, oversupplied market. They also continued to invest in refining and marketing operations despite the reluctance of oil-importing governments to authorize investments by state-owned enterprise. In such a market, however, long-term bilateral agreements between seller and buyer became crucial to the continued expansion of public oil enterprise.[49]

Initially disadvantaged in an asymmetrical situation, governments were able during the sixties and seventies to shift into balance-of-power bargaining. Matters were transformed, in sum, by an adjustment in the bargaining structure. Host governments changed the rules of the oil game by introducing a new institutional capability—the public oil enterprise.

But public oil entrepreneurs still needed political legitimacy and foreign finances from private-sector groups. Bureaucracies in Norway and Britain still relied on the political support of their parliaments; bureaucracies in Indonesia and Malaysia still required financial assistance from foreign investors. State bureaucracies relied less and less on societal groups for taxes, expertise, technology, and foreign exchange because of the obsolescing bargain. At the same time, however, domestic and multinational coalitions resorted more and more to other kinds of political and financial pressures to constrain state oil investments. They applied their influence more forcefully on representative bodies and central state finance agencies in order to extract concessions from public oil enterprises.

Pivotal Power Bargaining

Once balance-of-power bargaining between states and multinational and domestic companies replaced previous asymmetries, private actors sought new tactics to offset the growing power of public oil enterprises. One new bargaining principle was based on the notion that power is the ability to withhold small but critical amounts of resources upon which the state depends.

I call this kind of leverage *pivotal power*. This bargaining power exists when there is a balance of power among coalitions for and against a particular decision. The decision depends, or pivots, on minimal but critical resources—single votes or loans. In Norway the 14 percent capital financing for Statoil was the resource that the Norwegian Labor government needed during the 1970s to continue its oil expansion.[50] In Indonesia short-term loans in the 1970s for investments in fourteen oil tankers, and the unified backing of president and army for the state oil company, were the capital and political resources that the state needed to expand public oil enterprise internationally.[51] In neither case was the entire capital or power of the state or its oil company at stake. In both the resources at the margin of the state's ability to gain capital and retain political support were pivotal.

Evidence from Norway, Britain, Indonesia, and Malaysia shows that domestic and international groups were able to gain concessions from public oil enterprises by finding pivotal points of state vulnerability. Groups used their ability to withhold relatively small amounts of votes or finances in key state decisions on entrepreneurial oil investments or policy. Norway and Britain were more vulnerable to domestic political and financial threats to state power. Indonesia and Malaysia were more vulnerable to international threats of discontinued financing or delayed contract agreements. In such decision-making situations, government opportunities to pursue policies without private-sector support were scarce and costly.

Norway provides a dramatic example in the way that shipping posed major threats to the state through Parliament. Using their influence in the Conservative and Liberal parties, Norwegian shipowners created an impasse for Statoil by convincing Parliament to restrict to company's finances. Statoil requested in 1976 that Parliament grant it share capital to finance its development of the Statfjord oil field. Parliament, which controlled the company's budget increases, responded by cutting Statoil's request by 14 percent. Threatened in its entrepreneurial pursuits, the government was forced to negotiate an informal trade-off involving the Ministry of Industry, Statoil, and Saga, the shipowners' oil company.[52]

This example shows how the vulnerability of state interests to particular parliamentary finances jeopardized the Norwegian government's effort to pursue oil and exclude other domestic industries. Initially the Norwegian state was strong enough to achieve its oil goals autonomously. But soon domestic shipping groups with power in Parliament were able to bend state actions to their own interests. However, the state never abandoned its primary goals: a

fully state-owned oil enterprise, a government-guided national oil industry, and a majority share in that portion of oil operations owned by Norwegian companies.

If domestic interests curbed the state in Norway, in Indonesia international groups had more clout. The public oil company, Pertamina, was clearly trying to wrest control of the Indonesian oil industry and a share of marketing operations from the Majors. In 1974–75 an opportunity appeared for these international groups to ally themselves with economists inside the government to question Pertamina's extensive oil investments—particularly those involving international shipments of Indonesian oil. In 1973 Pertamina ordered fourteen ocean-going tankers to transport its oil supply, then 800,000 barrels per day. This was the final step necessary to create a fully integrated public oil company at the international level, where the multinational oil companies had previously held exclusive management control. But the step also provided foreign interests with an opportunity to threaten the company. At that time Pertamina was carrying a debt of nearly $10 billion from its investments, and in 1974 foreign banks called for debt payments on the short-term loans that were keeping the company afloat. Pertamina defaulted, and the threat of bankruptcy brought the company to its knees. President Suharto and the army, both of which had supported Pertamina's oil policies up to that point, were forced to side with government economists who opposed the public oil company.[53]

My unexpected finding, in Norway and Indonesia in particular, is that strong public oil entrepreneurs were susceptible to the deployment of small but pivotal bargaining chips. Private domestic and international groups were able to influence state oil investments and policy by withholding the resources upon which governments depended despite their broader independence of action. Norwegian shipowners and foreign banks were able to force the state to bargain on their terms by withholding small but critical amounts of finance. Chapter 3 discusses how the same was true in Norway and Britain for marginal or "swing" parliamentary votes controlled by fishing communities. The absence of these votes or finances would have jeopardized the state's leadership or survival as an entrepreneur.

Several kinds of state bodies in the four countries were susceptible to the withholding of pivotal political support or finance. Representative branches of the state were obvious channels through which societal groups could influence state policy: coalitions that had majorities in parliament by only small margins depended upon the electoral and parliamentary votes of their constituencies in order to retain their majorities and push specific policies. But other state agencies involved in financial or political agreements with societal groups were also potential channels for influence by private groups. Ministries of finance and public oil companies themselves depended upon continued external financing from international banks, foreign companies, or foreign governments. Public oil companies were also vulnerable if they engaged in business deals so important that the costs of delayed contracts could place entire governments in jeopardy. Finally, foreign ministries that wanted to maintain diplomatic ties with particular foreign states, such as the United States or Japan, could be subjected to intense outside pressure.

Pivotal power may be more important than any single type of balance-of-power bargaining leverage. The absolute amount of influence that one actor has over another by virtue of the resources it controls is not critical. The key, rather, is the relative share of influence that actors can exert at the margin by virtue of the particular votes, diplomatic support, or finances they control. By implication, it does not matter how much power a group has. What does matter is how the group uses the power it has to shape the decisions of the state.

The result of the application of pivotal bargaining power was that private societal groups—both domestic and international— were able to gain concessions of operational rights or ownership from the state, concessions that produced compromises between public and private sectors in the oil industry. The public appropriation of financial and industrial gains from the exploitation of national resources was transformed into a compromise with the private sector. Outcomes in developed countries reflected the demands of domestic sectors more than did those in less developed countries, where outcomes represented concessions to international groups. Pivotal-power bargaining thus forced governments to represent both their own autonomous bureaucratic interests in national and public control of oil and societal interests in private control of oil and shares or payoffs for such nonoil sectors as fishing and shipping. It may have meant a reduction in the exclusiveness of national public control and management of oil through public enterprise. But it also increased the extent to which both the bureaucracy's public interests and the society's private interests were represented in national policy decisions.

The shift to a balance-of-power relationship between state and societal groups may be the basis for more pervasive bargaining that relies on society's pivotal power. This is a change from the historical dominance of societal groups. The emergence of an autonomous role for the state as an entrepreneur during oil industrialization, it is clear, did not eliminate the state-society relations established during previous kinds of industrialization. Representative, democratic forms of state authority in early industrializers such as Britain insured that private domestic groups would be able to oppose actions of the "autonomous" state by withholding votes or finances in state bodies (for instance, parliaments). In later industrializers, such as Indonesia, states may have been highly centralized but they still depended upon multinational companies and international banks to sustain their oil production and financing. International groups could still bend entrepreneurial states to private investment purposes by withholding bank loans or cutting production levels.

The notion of pivotal power in state bargaining is significant for several reasons. First, private groups can force states into bargains that further the interests of private-sector investment. They can create bargaining situations by withholding parliamentary votes and capital resources that bureaucracies might have taken for granted. Threats that endanger state power or finances have forced states to strike bargains with domestic and international groups. These bargains often involve restricting public enterprise to certain industry operations, giving competitors preferences or simply leaving them exclusive market share, and granting actors special industrial rights or compensatory benefits for being displaced by state activities. So have opponents of the autonomous state been drawn into or conceded a separate place in the national process of oil industrialization.

Second, pivotal power is important because it introduces distinctive characteristics into the bargaining between state and society. The resource exchange—not the bargaining structure or process—may be asymmetrical. Societal groups, whether domestic or international, may withhold small amount of votes or finances that they then give the state in exchange for much larger amounts of oil leases or contracts. Similarly, the value of the resources exchanged may be asymmetrical. This results from a bargaining situation initiated by a threat rather than by mutual interest. The absolute political influence of the threatening group may itself be minor; but as long as a balance of power exists, the group can act pivotally, swaying a decision one way or another because it controls the swing vote. Another characteristic of pivotal bargaining power is that in developed countries it may involve votes (or finances that are voted upon) which are central to the democratic political system. In the more centralized political contexts of less developed countries, finances may be more critical, more pivotal, than votes.

The third and final reason why pivotal power is important in state-society bargaining is that it has significant implications for the future of democracy and capitalism in national economies. Pivotal power can preserve the private sector's role in accumulating wealth and using democratic channels to influence policy. Domestic groups can use decentralized democratic channels in developed countries to withhold key resources if the state does not act partly in the interests of private investors. Coalitions of international and domestic groups can use centralized financial channels in less developed countries to withhold capital unless the state concedes some private-sector demands. This situation in less developed countries is not democratic, however, for two reasons. First, the state represents non-national groups better than it does national groups. Fishermen, for instance, may not be an organized group within the governing coalition; if they are, the coalition most likely will not represent their interests over the interests of foreign oil companies. Second, formally decentralized, representative state institutions such as parliaments are not effectively democratic in that comprehensive representation of the public is not achieved through voting by the electorate.

Domestic and Multinational Concession Patterns

The second bargaining principle is that different concessions can be extracted from the state depending upon whether coalitions are composed of domestic or multinational actors. Although we recognize this idea intuitively, I systematically demonstrate in Chapter 5 how bargaining in eleven countries falls into two main groups, developed and less developed countries. In developed countries both domestic and multinational coalitions opposed state oil enterprise. In less developed countries it was primarily multinational coalitions that opposed state enterprise.

Two main patterns appear in the bargaining concessions made by states. The first results from the strength that domestic opposition groups can muster to demand concessions from the state. Such groups are private industrial companies, labor unions, environmental or interest groups, and political parties or lobbies. Concessions may include industrial contracts, financial payoffs, delays in oil operations, or preferential treatment. Such groups almost exclusively seek concessions from economic activities within national boundaries. Typically, domestic coalitions extract such concessions by using societally representative bodies of the state (parliaments) to engage in pivotal-power bargaining. The pivotal resource withheld from the state may be either political votes or the finances needed to sustain or expand state oil enterprise.

The second pattern results from the strength that multinational opposition groups can muster to demand state concessions. These groups are coalitions of multinational oil companies, international banks, and foreign governments. In contrast to domestic groups, they demand preferential or even exclusive control of international oil marketing or transport operations. They also insist on particular types of equity control of industrial operations or control of oil exports. Seldom, however, do they accept purely financial payoffs. Multinational companies are generally more willing to concede national oil operations to the state in exchange for preferential or exclusive control of international operations, which distinguishes multinational concessions from domestic ones. Multinational coalitions rely on indirect pressure exerted on autonomous bodies of the state, such as ministries of finance, to extract concessions. As a result, their use of pivotal bargaining power is aimed primarily at withholding critical loans or loan rescheduling, tax payments, or contract agreements from state oil enterprises. They can seldom mobilize parliamentary pressure—unless the coalition also embraces important constituencies in domestic politics.

This difference between the state's domestic and multinational concessions is attributable to the relative dependence of domestic companies on home markets. Multinational oil companies remained sufficiently profitable up to the mid-1980s to be flexible negotiators. As they were forced by strong states to concede national oil operations, they recentered their profits in international marketing and transportation. Domestic companies, however, depended on profits from national industrial markets. They did not necessarily have overseas operations or sales through which to recoup national losses. Their only option was to play tough in bargaining with the state for oil-related contracts in areas such as the North Sea.

The countries analyzed in Chapter 5 fall into two main groups. The first group—Mexico, Saudi Arabia, Iran, Indonesia, Malaysia, and Italy—display variations on the multinational concession pattern. Despite opposition from multinational coalitions, state oil companies generally became fully integrated into national and international oil operations. The second group of countries—the United States, Britain, Japan, France, and Norway—shows how much states conceded when both multinational and domestic coalitions demanded concessions. Either state oil enterprises were slowed in their emergence (Japan), or they were entirely (the United States) or partially (France, Britain) eliminated in the face of societal opposition.

This book explains the evidence of state oil enterprise from a statist perspective. I concentrate on the Norwegian, British, Indonesian, and Malaysian cases. Chapter 5 then extrapolates the argument to seven other countries, including the United States. A statist perspective focuses on the interests, role, and actions of the state as a central actor in international politics and the global economy. States have previously been classified as strong or weak, according to their ability to guide societal coalitions, but I expand and modify this perspective. In this chapter I have discussed three ambiguities in the statist approach. First, I clearly differentiate state interests from societal interests by positing that bureaucratic autonomy is the source of independent state interests. Second, I believe we may argue that the autonomous state is a strong and at least partly effective actor based on the degree of continuity between state interests and state entrepreneurial actions. Third, pivotal power and different state concessions to domestic and multinational coalitions are, I have argued, distinctive bargaining principles across the four countries. This gives us reason to believe that a comparative statist logic of balance-of-power bargaining may allow us to understand, and potentially to predict, outcomes for state oil enterprises in many capitalist countries.

The next chapter examines why governments chose to resort to state-owned oil enterprise rather than stick to traditional taxation, leasing, and regulatory policies. Why were governments dissatisfied with their conventional capitalist roles, leaving oil production to private-sector companies? The four governments eventually resorted to policy packages that combined taxation, leasing, regulation, and state enterprise.

CHAPTER THREE

Oil Policy Options

Governments in Norway, Britain, Indonesia, and Malaysia could have been content with traditional oil policies. They could have filled their purses by simply demanding greater corporate income taxes, royalties, and bonuses on the production of private oil companies. They could have stockpiled domestic oil by taking their share of taxes in crude rather than in cash. They could have insisted that their citizens be given oil jobs by writing employment and local contracting requirements into lease agreements. And they could have protected their territory and their fishing industries by imposing environmental and safety regulations on oil drilling.

But those shrewd sovereigns insisted on trying their hand at the oil business. Was this stubbornness—because they could not squeeze enough from traditional oil policies? Or did private oil entrepreneurs defy their demands for national oil development? This chapter outlines the basic choices in tax revenue, management, regulation, and public ownership that governments could have made in formulating national oil industry policies. I then discuss case histories of actual choices in the four countries and compare them in the period between 1968 and 1985.

POLICY CHOICES

Governments had a wide range of types and degrees of policy control they could exert over oil industry operations, among them taxation, production ceilings, lease requirements, operational safety regulations, and public enterprise. Each could limit or subtract from private revenue, management, operations, and ownership control in different ways and to different degrees.

Revenue Control. Taxation, for instance, subtracts from the revenue control that a private company can maintain over the financial proceeds from its production of oil. The company retains complete control over the ownership, management, and operation of production, but it agrees to pay a share of income from the sale of oil supplies to the government rather than insist that it treat all revenues as direct profit or reinvested earnings. Taxation of revenues can take the form of area fees for exploration. It also includes information and production cash bonuses paid when lease agreements are signed or during production as well as corporate income taxes and royalties paid on production. Royalties can be shifted from a fixed to a sliding percentage to speed up exploration or production by making immediate operations less costly than delayed ones. In addition a variable royalty enables a government to benefit from price changes that alter the value of the oil upon which the royalty is calculated. It has been implemented by governments in Thailand and Brunei.[1]

As a group, these forms of taxation constitute part of the economic rent generated by the production market. Theoretically, the amount that governments receive in rent equals the company's total oil revenues minus its total costs, which include a share taken as profits. Clearly, the greater the profit demanded by the company, the less will return to the government in economic rent. The limit on how much governments can increase their rent share depends upon the smallest profit share that private companies will accept before they declare oilfields worthless to exploit commercially or threaten to leave and produce oil elsewhere. The greater the competition among private companies for access to oilfields, the smaller will be the profit share they are willing to accept in order to gain access. The ability to waive taxes or create moratoria on payment is a bargaining chip for the government; the ability to withhold tax payments is a valuable chip for private companies.[2]

Profit reductions reduce a company's ability to reinvest earnings in oil operations in the same or another country, unless reinvestment precedes the declaration of profits. Had the Norwegian government siphoned off a larger share of Norwegian shipowners' profits in the early 1970s, for example, Norwegian shipping companies would have been less able to form an oil company and diversify their investments into Norwegian oil production. Multinational companies sometimes avoid profit losses from taxation by transfer pricing. This accounting procedure enables them to have their profits taxed in countries with relatively low taxation rates. Companies then write off losses in countries with higher rates.

Another form of revenue control by states is the ability to siphon external borrowing from banks into public operations. This practice reduces private-sector control over oil industry investment by funneling private capital (in the form of loans) into publicly rather than privately owned operations. The creation of a large public debt thus gives the state revenue control by expanding its investment of capital resources. But a large debt also increases bank control over public finance. Banks can refuse to renegotiate payment schedules if debtor governments become negligent in servicing loans. The Indonesian state oil company accrued a $10 billion debt from independent banks during the early 1970s, enabling it to become one of the world's largest oil companies. However, it was the company's negligent servicing of those loans, and the unwillingness of banks to renegotiate payment schedules, that eventually led the company to near bankruptcy and a corporate reorganization.

Management Control. Second, governments can assert policy control by interfering in the private management of oil operations. Management control can take several forms. Governments can set a limit on total national oil production. In the 1970s, for instance, the Norwegian government established a production ceiling of 90 million tonnes of oil equivalents per year. This ceiling insured that the production rate would neither deplete the resource nor force the country to absorb oil revenues and oil-related changes in the economy at too quick a pace. Such production limits constrain total national production, but they do not generally interfere with production levels in individual companies. The number of potential production sites ("blocks") that a government decides to open for bid determines whether the production ceiling will be exceeded or not. Companies can produce at maximum rates as long as the total number of producing wells is limited to guarantee that the national production ceiling will not be surpassed. Choice of leasing policy is thus critical to the implementation of a production ceiling.

Once governments determine the number of blocks for which to invite bids, they have another decision to make. They can choose oil companies based on competitive bidding that maximizes the rate of extraction and government rent. Or they can choose oil companies based on administrative bidding. Administrative bidding enables governments to require the employment of nationals and local businesses and performance in exploration and production according to specified schedules. The government then chooses the company that best satisfies its requirements and schedules. This form of bidding also gives governments relatively greater ability to discriminate among companies for political or nationalistic reasons. Once Norwegian and British governments required that national labor and industries be used offshore whenever possible, in the mid-1970s, they ended the freedom of foreign oil companies to bring in low-wage laborers from Spain, Portugal, and North Africa and to hire U.S. contractors.

Management control through production limits, leasing requirements, and bidding practices thus enables governments to assert more control over oil operations than they can through taxation. Governments can expand their share of revenues by increasing the total number of oilfields generating taxable profits. Private companies prefer this situation to the loss through increased taxation of a greater share of potential profits from existing fields.

Operational Control. Performance and safety regulations provide a third way for the government to exert control. Such regulations subtract from the operational control of private oil companies by reducing their freedom to determine where and when specific drilling or production will occur, how tankers will be maneuvered, and so on. Governments in Britain and Norway required, for instance, that oil companies drill offshore only during "safe" weather conditions, to avoid oil spills. They also insisted that companies clean up their oil drilling debris from the sea floor to reduce oil-related mortality of fish and damage to fishing gear. But operational control by government also worked in favor of private oil companies. The Indonesian government has used prohibited zones to keep fishing operations from physically interfering with oil operations, and in 1978 fishermen were required to avoid seven offshore areas of six hundred square miles each.

As sovereigns, in sum, governments can carve out a public share in national oil industrialization. They can collect oil revenues through taxes, assert management control through leases, and claim operational control through regulations. In these ways they govern oil activities at the level of the nation but do not interfere directly in oil business at the level of the firm.

But if governments become dissatisfied with their share as sovereigns, they may begin to take a more assertive role at the level of the firm. They can gain a larger share for themselves independent of society if they get directly involved in the corporate production of oil and gas rather than leaving it exclusively to others. Entrepreneurial governments have varied greatly in the degree of public control they assert through public participation and, ultimately, ownership.

Ownership Control. Ownership control can reduce the relative market share that private companies enjoy in production leases and contracts for transportation, refining, distribution, and subcontracting (including supply boats and drilling rigs). Other than concession agreements, which involve no government ownership, most production agreements carve out some form of equity participation for national governments. Such agreements include profit-sharing, carried-interest, joint-venture, production-sharing, and service contracts.

Profit-sharing contracts were used first in Venezuela in 1948 and later in the Middle East. Such contracts secure a direct share of oil company profits for governments. In Norway a variation on profit sharing was used briefly during the 1969–70 licensing round to try out government participation on a net-profit basis.

The unique feature of carried-interest agreements is that governments can automatically gain a share of equity and then pay for it out of their contractual proportion of production earnings. Norway began to use carried-interest agreements in 1969 and continued to use them as the major form of government participation through the 1980s. Participation started either the moment the license was granted or once a commercial discovery was made. Once a discovery was declared commercial, the state oil company assumed ownership of a share of up to 90 percent of oil profits; the percentage depended upon the size of the field. At the same time the state company took a share of management control equal to the profit share and after ten years was entitled to take over full operational control of oil production from the field. Costs of production were also shared between private and state oil companies, although exportation costs were covered entirely by the private company.[3]

Joint ventures are a third form of equity agreement. Governments first create their own state oil companies and then form business partnerships with private oil companies. Theoretically, profits, costs, management, and operations are shared, as in any joint-venture contract among private parties. Although joint ventures were an option in Norway, they did not gain the popularity of carried-interest agreements because of the immediate cost-sharing requirement. The Malaysian government, however, did agree to joint-venture contracts to gain equity participation in a new refinery built by Shell in the 1970s. It also used joint ventures to participate with foreign companies in the formation of new exploration and production companies during the 1980s.

Production-sharing contracts are commonly used in the South China Sea. They were also an untried official option in the North Sea.[4] Initially developed in Indonesia in 1966–67, production-sharing contracts gave governments ownership of oil supplies before sale. This was a major shift from earlier profit-sharing agreements that divided only the after-sale profits between governments and multinationals. Presale shares reduced currency-related problems and gave governments the chance to sell their oil at prices and to customers of their choice. Indonesia used this new freedom to break into the Japanese domestic market in the early 1970s, when Pertamina collaborated with independent Japanese oil interests rather than the Majors.

Production-sharing contracts not only shared control of oil supplies, they also shared costs and profits, but only after production. All risk costs for investment and development were paid by private companies. After production had commenced, governments enabled the multinationals to recover their expenditures and operating costs out of a portion of production called "cost oil." The remaining portion, "profit oil," was divided between the government and the company.[5] In Indonesia the original cost oil proportion was 40 percent and the initial profit oil split was 65–35 in favor of the government (including the state oil company, Pertamina).[6] As windfall profits began to accrue to multinationals with the price rises of the 1970s, however, governments demanded that both cost oil and profit oil be calculated using a sliding scale based on level of production and current oil price. Governments also used royalties or cash bonuses paid upon the signing of leases or upon commencement of production to capture a greater share of oil company profits.

The final component of production-sharing contracts was that governments retained full ownership and management of oil operations and supplies. In practice management was sometimes shifted to private contractors, to facilitate production. Official management and government ownership ended, however, once supplies were exported. At that point the multinationals gained full ownership and management of their contracted share according to the profit oil formula. Such production-sharing contracts served government interests in increasing ownership, management, revenues, and operational control over oil production operations. However, stringent control on revenues was a disincentive to multinationals to explore or report small, high-cost oilfields.[7]

Finally, service contracts give governments the chance officially to retain ownership of oil supplies and full control of the production lease. In practical terms service contracts are often little more than new labels for old concession systems, except that profit shares may change. The work contracts instigated in Indonesia in 1963, for instance, were a version of service contract that simply sustained Caltex's and Stanvac's earlier concession agreements. In a typical service contract, private companies may retain full control over profit shares, costs, management, and oil supplies, which have to be purchased from the government at market prices.

These five types of agreement vary over whether they split profits, costs, management, and ownership of the oil supplies between the government and the private company. The typical case for each one is indicated in Table 3.1. Although there are legal distinctions

Table 3.1. Types of government participation agreements in oil
production operations

Type of control	Type of agreement				
	Profit-sharing	Carried-interest	Joint-venture	Production-sharing	Service-contract
Profits	G/P	G/P	G/P (or fee)	G/P	P
Costs	P	G/P	G/P	P	P
Managerial	P	G/P	P	G (operations = P)	P
Oil supplies	P	G/P	G/P (if not profits)	G/P	Gª

ªPrivate company has option to purchase at market price.

Note. G = government share; P = private company share; G/P = a split of shares between the parties.

Source. Information from Roderick O'Brien, *South China Sea Oil: Two Problems of Ownership and Development*, Institute of Southeast Asian Studies, Occasional Paper no. 47 (Singapore, August 1977), pp. 51–73.

between joint-venture, production-sharing, and service-contract agreements, Roderick O'Brien and Charles Johnson have pointed out that the practical commercial effects are the same, because of the competitiveness of the oil environment for both companies and governments. Agreements are unlikely to have terms of risk, product, and cash that differ from those which companies can conclude in another country.[8]

But from the plethora of taxation, leasing, regulation, and participation options, which policy packages did governments in Norway, Britain, Indonesia, and Malaysia actually choose to achieve their national oil development aims? Did private oil companies already at work welcome or balk at these policy choices?

Actual policy choices in Norway, Britain, Indonesia, and Malaysia between 1968 and 1985 show that governments favored state ownership of oil production. All four governments initially claimed public shares of oil revenues, industrial management, and operational control. They increased rates of taxation, required the use of national companies and employment as conditions for gaining leases, and defended other domestic industries by regulating oil

75

operations. But ultimately all four relied on state ownership of oil and gas production, transportation, refining, or distribution to pursue national oil objectives. The Indonesian and Malaysian governments gained a larger share of equity, management, and revenue control than did the Norwegian and British governments.

NORWEGIAN OIL POLICIES

At the start of North Sea oil and gas expoitation, in the 1960s, the Norwegian government was content to rely on its taxation, leasing, and regulatory powers as a sovereign. The 1958 Geneva Convention on the Continental Shelf had given governments the exclusive international right to exploit natural resources on their continental shelves. By 1965 the Norwegian and British governments had signed an agreement "Relating to the Delimitation of the North Sea between the Two Countries."[9] This agreement enabled both governments, free of international dispute over boundaries, to grant leases to oil companies to explore, develop, and produce their newly acquired national oil and gas resources in the North Sea.

Like most other governments, the Norwegian state initially calculated that the investment capital, vertically integrated operations, and oil experience of foreign multinational companies would provide the most efficient exploitation of Norwegian oil and gas. Besides, except for the Norwegian chemical company Norsk Hydro, of which the state owned 48 percent, Norway lacked nationally owned capacity in the oil industry. So between 1966 and 1977 Phillips Petroleum, Esso, Mobil, and Petronord (involving Total, Elf, ERAP, and Aquitaine, as well as Norsk Hydro) started working in Norway. They funneled new North Sea supplies into their existing global networks of oil transportation, refining, and marketing.[10]

The Norwegian government did begin to assert management control over foreign oil operations through its leasing policy. By choosing to administer licenses itself rather than auction them off to the highest bidder, the government used its political discretion to select those companies which would favor the government's microeconomic aims. Bureaucratic discretion was more important than gaining a maximum share of economic rent by bidding up the lease price. The Ministry of Industry was able to encourage the exploration of less attractive areas by smaller companies and to insist that foreign companies employ Norwegian goods and services as a condition for license approval.[11]

The government also used leasing policies to speed up oil extraction. By dividing offshore areas into blocks rather than large lease areas, the government stimulated companies to explore and develop smaller areas more quickly. Also, in 1972 the government reduced concession periods for licenses to thirty-six years, pressuring companies to develop oilfields in substantially less time than the sixty- to ninety-two-year periods typical in the Middle East.

Finally, the Norwegian government asserted managerial control over private operations by setting a ceiling on the rate of oil and gas extraction. In 1973 it imposed a yearly production ceiling of 90 million tonnes of oil equivalents. The government then granted new licenses in accordance with a "moderate" rate of national resource depletion aimed at keeping Norway a net exporter until the year 2000. By choosing a slower, more conservative pace, the government reduced the relative influence that private oil companies might otherwise have exerted over production policy.

But the government was more cautious in carving out a share of public oil revenues through taxation. Uncertain about its actual oil and gas reserves, the government hesitated to tax companies too heavily and thus kill their appetite for Norwegian oil. Between 1965 and 1972 the government collected a public revenue share composed of royalties, area fees, and taxes on the posted prices that OPEC had established above market prices to assure minimum revenues for producer governments. The initial taxation rate in 1965, 10 percent for oil and 12.5 percent for gas, was changed to a sliding scale of 8–16 percent for oil by 1972. Area fees of $70 per square kilometer for the first six years were charged in 1965, with a ceiling afterward of $700. Corporate income taxes were set at 45–50 percent of net income. However, transfer pricing and lower corporate income taxes for offshore operations reduced actual taxation rates into the 1970s.

After 1973 the government became more concerned about getting its "fair share" of the windfall profits accruing to the multinational oil companies because of OPEC price increases. In 1974 and 1975 the government reached agreement with the companies that they would be taxed at a substantially higher rate of 70 percent—which still gave them a 20 percent rate of return on their private capital investment. In 1975 the government also created a new excess profits tax of 25 percent—the Special Tax—on residual profit after income taxes and royalties. A new pricing system accompanied the tax, in which the government established the "norm price," a notion of the "real" market price of crude traded on a free market by independent traders.

But the Norwegian government was generally frustrated by the minimal economic control that leasing and taxation policies provided. It decided to shift its development approach from "hands off" to "hands on." First, the government took a 51 percent, controlling interest in Norsk Hydro. That company was producing 5 percent of the Norwegian share of the Frigg gas field, jointly owned by Norway and Britain. The Norwegian government downplayed Norsk Hydro's role, however once it had reassessed its share of incoming revenues and sought other channels of state involvement offshore. In 1969–70 the government introduced state profit-sharing participation in leases to foreign oil companies. The participation percentage was based on the net profits of foreign oil companies. State participation increased from 5 percent in the Frigg field in 1969 to 40 percent in the Heimdal field in 1971. The government also introduced carried-interest agreements during this period; these would become popular after 1972.[12]

By the early 1970s it was clear that the state's interests in increasing its economic control through oil and gas development were not being realized. Owing to various cost deductions for developments and reductions in tax yield from capital allowances, revenues from taxes and royalties provided the government with only a 20 percent share of total returns from oil production, a much smaller share than OPEC countries received. In 1972 total gross product from production was $31.3 million, but the central government's net income was only $6.3 million. In 1971 Norwegian production of crude oil from seventeen operating wells was about six thousand barrels per day (b/d), 1.5 percent of West European output and an insignificant share of world production. Virtually all of the Norwegian oil was exported, though Norway's imports neared 200,000 b/d. The government wanted a larger share of revenues, and it wanted to halt the country's growing deficit in the oil trade by producing more and keeping more for domestic use.[13]

Statoil

Anxious to stake out a public claim to ownership of offshore oil production, the Norwegian government formed its own 100 percent state-owned oil company, Statoil, in 1972. State production, unlike arrangements with private Norwegian companies, would provide the government with direct income to supplement what it received from fees, royalties, and taxes. It would also give the state an opportunity to supply domestic oil needs directly as well as to buy oil from foreign companies. Besides, the state wanted "an opportunity of influencing the decisions made by the individual licensees. The Government will thus be better able to guide and control activities to the extent desirable and necessary to ensure that continental shelf resources benefit the Norwegian community as a whole."[14] Practically speaking, this meant that the government would increase its control over the link between national production and offshore services.

Rather than threaten foreign companies by retracting oil licenses, the Norwegian government established cooperative production arrangements with them, thus retaining foreign risk capital and know-how. Statoil participation was simply substituted for government participation after 1973.

The formation of Statoil was the government's first entrepreneurial policy. Statoil took the share allotted to the government in the previous state participation policy. By 1973 Statoil participation had increased to 50 percent in the Statfjord field, and by 1975 Statoil was granted 50 percent in all blocks conceded to private oil companies with the option to take up to 25 percent more. By 1976 the company controlled 50 percent of ten of the eighteen blocks licensed after 1971. However, Statoil's share of oil supplies was not commensurate with its share of production control. In 1976 the production share of foreign oil companies was about 270,000 b/d;

Statoil's production share was only about 3,500 b/d. Furthermore, Statoil agreed to distribute its supplies within the domestic market, leaving foreign oil companies in undisturbed control of oil exports.[15]

The Ministry of Industry also proposed to the Norwegian Parliament that Statoil earn additional profits and coordinate Norwegian oil industry development by integrating vertically while continuing to cooperate with private business interests. Vertical integration would give it a central role in guiding the growth of related industries:

> Such a company will also give the Government a better opportunity to take advantage of the rights acquired under the contracts relating to government participation as a basis for engaging in transport, refining and marketing. The company's other objective will be to coordinate the Government's interests with the interests of private businesses within the petroleum sector through various types of cooperation agreements. The company would thereby play a major part in the realizing the Government's policy of establishing an integrated Norwegian petroleum community.[16]

Statoil moved rapidly into gas in 1973 by securing a 50 percent share in the pipeline-laying company for the Ekofisk field and committing half of Statoil's equity capital to the company.[17] The state oil company also had prospects for investments in shipping and other transportation.

Despite its formal recognition of other nations' oil companies, the Labor government worked to increased Statoil's monopoly over national participation in offshore production. Statoil was the "Labor party's baby." The government used it to assert national and public economic control of domestic oil production, but it also used it to serve its partisan interests in keeping Conservative party oil interests at bay politically. Before 1973 the Ministry of Industry claimed that no other private Norwegian company or group of companies had sufficient financial strength and experience to become operator for an entire block.[18]

The Labor government had several reasons for using its managerial control over licensing to favor a Statoil monopoly of the government's 50 percent share of production. Limiting contracts offshore for other Norwegian oil companies would keep domestic private-sector interests from gaining a strong market position in national oil. The Labor government's economic and political power would thus be secured as oil continued to develop as the single most important sector in the national economy. Oil revenues would not only provide oil funds for government programs but also give the government a new source of income and foreign exchange for the future. It would thus enable it to end its dependence on the private sector's critical role in providing foreign exchange to help balance international payments. In this way the Labor government could diminish the strength of the private sector in the national economy by turning a public company into the nation's strongest firm. Furthermore, a Statoil monopoly over the share of oil to be produced by domestic industry would provide a basis for later extending the state company's control through investments in national refining, transportation, and marketing.

Norwegian Shipping's Response

Alarmed by such possibilities, Norwegian industrialists focused on the economic and political importance of gaining their own share of North Sea oil profits. While the semiprivate Norsk Hydro continued to work within the Petronord group, producing gas from the Frigg field, other private Norwegian interests, including shipping, pooled capital to invest in oil and challenge government policy. Norwegian shipping companies, in particular, saw entry into oil production as a way to maintain their relative strength vis-à-vis government oil. Offshore oil investments would also give shipping companies vertically linked operations in oil production and transportation.

The shipping industry had important factors working in its favor as it struggled to enter the oil industry. Shipping had long been one of the country's most important and profitable earners of foreign exchange. It was also a Norwegian-owned industry: only about one-quarter of the fleet was foreign-owned, which gave it a legitimate domestic identity. Moreover, shipowners had already expanded into oil-related operations. Investments in the 1960s had made one-third of the Norwegian fleet oil tankers by 1970. Shortages on the world tanker market brought soaring tanker profits in 1970 and 1971, providing Norwegian shipowners with capital to reinvest. As a result, fifty shipping companies and forty companies from other industries joined to form their own oil company, Saga Petroleum, in 1972.[19]

The government's preferential treatment for Statoil limited Saga's participation in Norwegian licenses to only 8 percent of one block and 2 percent of two others by 1973. In comparison, Statoil received 50 percent of the Statfjord field.[20] Moreover, Statoil provoked shipowners by advancing concrete proposals to invest in shipping and other domestic industries linked to oil production.

These entrepreneurial state policies incited private industry to mobilize opposition in Parliament. Using their influence through the Conservative and Liberal parties, shipowners created an impasse for Statoil. Parliament controlled the state oil company's budget, and in 1976, as already noted, it cut by 14 percent Statoil's request for share capital to finance its Statfjord development.[21]

By mobilizing this parliamentary threat to Statoil's funds, Norwegian shipping interests forced the government to compromise on state investments in oil-related markets, offshore and onshore. The government conceded control of domestic subcontracting operations, such as supply boats and drilling rigs related to oil production, to private shipping interests, but it insisted that Statoil retain its monopoly over the government share of production. This trade-off was made informally between government oil interests, working through the Ministry of Industry and Statoil, and Conservative party shipping interests in Saga. Previously Exxon, Phillips, and other operators had subcontracted work to other foreign companies. Because by 1975 Statoil participation offshore was up to 50 percent in many fields, however, the company could influence a corresponding share of subcontracting decisions.[22]

In addition, the government began to enforce 1972 National Continental Shelf regulations. The "full and fair competition" clauses in the regulations (Decree 8.12.72, sec. 54) stated that oil operators should use private Norwegian goods and services whenever competitive. Starting in 1975 the portion of offshore subcontracts held by private Norwegian companies (mostly shipping companies) began to increase steadily—29 percent in 1975, 41 percent in 1976, and 55 percent in 1977. By 1979 Norwegian subcontracts represented up to 60 percent of the total for offshore drilling and supply boats. If we combine drilling rig and supply boat investments, the total value of the Norwegian offshore fleet owned by shipping companies in 1978 was about $2 billion.[23] So although Norwegian shipping did not gain a larger share of oil production, it did get a substantial share of subcontracting.

The government also assured Norwegian shipowners that state companies would not invest in shipping-related operations.[24] It would thus preserve Norwegian shipping's share of the transportation of Norwegian oil in national flag tankers. Of course, this agreement said nothing about oil pipelines, in which the government later did become actively involved to the detriment of tanker transportation.

The government may have succeeded in retaining the Norwegian share of oil production for Statoil until 1978, but that share was small in actual oil supplies. In 1976 the company's production was only about 3,500 b/d for domestic use, compared to the 272,000 b/d produced by foreign multinational oil companies. The net-profit and carried-interest production agreements with Statoil gave foreign oil companies continued control over almost all Norwegian crude exported to other countries. In 1981 Statoil crude production was still only about 14 percent of total Norwegian production: 58,000 b/d compared to the 414,000 b/d produced mostly by foreign oil companies. By 1986 Statoil had surged ahead to produce nearly half of Norwegian oil: 320,000 b/d of the 1985 total of 786,000 b/d.[25]

But by the late 1970s Norwegian policies had shifted, in response to OECD efforts to break dependence on OPEC oil and find substitute energy sources. In gas production offshore the Norwegian government began favoring domestic participation rather than just control over marketing. By 1980 a conflict had arisen between the Norwegian and U.S. governments as Norway pressed for parity between gas and oil prices. Norway argued that parity was necessary for Norwegian companies to enter the market and exploit gas resources profitably, while the United States arged that it was neither necessary nor wise.[26] Less worried about parity, the multinational oil companies opposed the concession of any larger share of national gas production to domestic interests. By 1981 the trend seemed to be toward nonparity prices but assured supply. Norwegian gas prices were nearing world market prices (on a delivered basis) for gas entering the new Norwegian North Sea gas-gathering system.[27]

Though the Conservatives took control of the Norwegian government in 1981, there was little real change in Norway's oil and gas policy. Strong political opposition from Norwegian trade unions thwarted initial efforts to spin off Statoil's equity shares of operations to the private sector. The unions wanted to preserve public control of oil and with it their own political influence within the industry. They forced Conservative leaders to stick to the state's prior policies in the pursuit of national and public goals in oil and not be diverted by partisan persuasions. The Conservatives succeeded in increasing private-interest shares only by expanding Saga and Norsk Hydro participation relative to that of Statoil and by making possible the participation of new Norwegian companies as more northerly fields opened. Norsk Hydro, for instance, was doing significantly better by January 1980 than it had in 1978, with profits from Norwegian oil operations up 60 percent. It also held shares in at least eight fields under production or exploration. The company's output from its Phillips group share and its 20 percent share of the Frigg field alone gave it a total output equivalent of 12.3 million tonnes by 1980. However, the overall share for Norsk Hydro was still significantly less than Statoil's.[28]

Norwegian Fishing's Response

Norwegian policy changed in the mid-1970s for another reason: other domestic industries were unable to compete for space with oil in offshore markets. By 1970, 40 percent of the Norwegian fish catch was coming from the North Sea, but about 50 percent of the Norwegian sector south of the sixty-second parallel was in various stages of oil and gas exploitation. Fishing suffered from operational interference: nets were damaged and navigation was impeded by oil activities. The Directorate of Fisheries in Bergen estimated that access had been reduced between 15 and 85 percent in various major fishing grounds. Government and fishing industry sources independently estimated fishing vessel losses of between 30 and 40 percent in catches of certain stocks during 1975–77.[29]

By 1975–76 southern fishermen had responded by organizing politically. To reduce this opposition to oil development, the government began granting compensation for gear damage through the Directorate of Fisheries. The total amount granted by August 1977 was $1.67 million.[30] Compensation was not given for loss of space because fishing is legally a public right, not subject to private claims of property offshore.

North of the sixty-second parallel, however, the Norwegian fishing industry and environmentalists temporarily halted oil development. In 1971 the government established that parallel as the northernmost limit for oil licensing because Norway, Britain, and the USSR disagreed on how to divide rights on the continental shelf above that line. As spatial conflict between fishing and oil increased south of the sixty-second parallel, however, fishing interests north of the line joined environmentalists and liberals lobbying in Parliament to prevent future northern oil operations. Supported by the Agrarian, Christian, Liberal, and Socialist parties, this "Green Opposition" called attention to oil spills, damage to the fishing industry, and changes in the economy and lifestyle in the region. By contrast oilmen, backed by the Conservative party and a coalition of the construction, shipbuilding, and shipping industries, plus southern labor union groups, pressed for northern licensing. They argued that more drilling would increase the rate of oil production and economic growth, provide jobs, and increase government revenues.

The sixty-second parallel became the battle line, Parliament the battleground. Political debates raged over when to start northern drilling. Northern Norway's fishing had pivotal political clout because fishing communities controlled two seats critical to the Labor party's parliamentary majority. These fishing interests threatened to shift the Socialist votes of northern fishermen if the Labor government, which was supported by the Socialists in Parliament, allowed drilling north of the sixty-second parallel. But in 1977 the Labor organizations of two northern fishing countries, Troms and Finnmark, decided to support northern drilling because oil com-

panies promised to develop oil-related support and service industry onshore. This shift of parliamentary support unbalanced the
northern stronghold. Political debate diminished, and the government's environmental position eased. Parliamentary discussions
came to focus on the operational issues of safe drilling procedures,
and by early 1978 opinion in Parliament had grown favorable to
starting northern drilling in 1982.[31]

Three factors continued to favor fishing and environmentalist
interests. Oil spills and platform destruction in rough North Sea
weather delayed northern drilling plans for environmental reasons. In addition, higher oil prices made previously uncommercial
southern fields look very profitable, relaxing the pressure to develop in the north.[32] Finally, northern capelin fishing continued to be
very profitable, accounting for one-quarter of the total value of the
Norwegian catch of $457 million in 1976.[33] So in the south the
government compensated fishermen, but in the north fishing temporarily generated a political threat to the Labor government
which altered state oil policies.[34] Nonetheless, some drilling was
under way in northern waters by early 1981.

Oil Boom and Bust

By the 1980s it was clear that the oil boom had created two
contradictory trends. Norway had become wealthier from oil revenues but also more indebted and industrially impoverished from
the investment requirements and inflationary effects of oil.

Oil revenues had been only $90 million in 1974; they were up to
$5.5 billion by 1981, largely due to the doubling of Norwegian oil
prices in 1979 after OPEC's price rises. Oil and gas income contributed about 80 percent of the gross national product in the same
year. Oil exports had also boosted Norway's trade position: they
constituted over 30 percent of the country's total foreign trade.
And by 1981 oil had converted the trade balance from a deficit into
a surplus.[35]

Other industrial sectors were hurt by this oil revenue and trade
boom, however. Before 1973 Norwegian industries had been little
affected by changes in the global oil market, because their imports
were negligible. But after 1973 Norwegian industries became less

competitive, because of oil-induced inflation and currency appreciation. Companies were near bankruptcy and industries were in decline in shipbuilding, fishing, farming, and textiles. Committed to full employment, the government heavily mortgaged future oil revenues in order to subsidize these industrial sectors—to a total of $14 billion between 1973 and 1980.[36]

The Norwegian government also used its management control through administrative leasing policies to help employ failing Norwegian industries. In 1979 it granted leases to those foreign oil companies which promised to invest in joint ventures, or other partnerships with Norwegian firms, that would strengthen employment and Norwegian industry.[37]

The price of oil wealth and industrial subsidies was increasing financial troubles for the government. Norway had started with a positive balance of payments before 1973, but now increasing costs and delays in oil and gas extraction forced the government to borrow money abroad. Foreign borrowing started in 1974, aimed at three objectives. First, more capital was needed to finance the exploitation of North Sea oil and gas, because of unanticipated costs and delays. Second, foreign loans were used to bail out an overexpanded Norwegian shipping industry, suffering from global decline in shipping after 1973. Third, subsidies for firms and industrial sectors negatively affected by the oil boom required more capital than could be taken from existing oil revenues. By 1977 foreign loans had contributed to a $22 billion debt. By 1979 that debt had risen to 27 billion. One-quarter of that 1979 debt was government borrowing related to oil.[38]

Despite these contradictory trends, Norway had substantially improved its overall position by the 1980s. The government had managed to lower the rate of inflation while modest economic growth continued. It had also sustained high employment, improved Norwegian industry's competitive position, and kept oil and gas revenues increasing. But though oil and gas income flowed, the nonoil national deficit continued to rise. Much of the oil income had to be used to offset the growing deficit.

The Norwegian government continued to pursue tight taxation policies regarding multinational corporations. A new draft tax law was made public in 1980 which increased the government's royalty from 16 to 18 percent. The intention was not to dissuade development activities but to increase revenue control through increased rates of taxation, which would supplement borrowing from overseas banks. The government also shifted the supervision of technical and business matters regarding oil production to the Ministry of Oil and Energy, increasing the stringency of government operational control. Furthermore, the new legislation enabled the government to force oil companies to relinquish oil concessions unless they made certain investments and to concede to requisition requirements if necessary to meet Norwegian oil supply agreements.[39]

In 1980 the government also reaffirmed its conservationist, "moderate-pace" production policy. But it also began to exceed the earlier ceiling of 90 million tonnes of oil equivalents. The government asserted, however, that despite production increases it would retain firm control over the timing of field developments and over the volume and distribution of oil and gas production. Between 1980 and 1982 production remained static at about 49 million tonnes of oil equivalents annually, but by 1983 it had risen to 54 million tonnes. By 1984 the government was launching a major effort to boost oil and gas output into the mid-1990s by 60 percent. In the absence of new oilfield discoveries, production was expected to peak by 1990 and then decline sharply. Falling international oil prices meant that the rate of production would have to increase to sustain government oil income.[40]

In this production surge Statoil retained its dominance over the Norwegian-owned share. Although the Conservative governing coalition continued attempts to increase parliamentary control over Statoil, the state oil company remained intact. Its survival was championed by a former Labor leader committed to entrepreneurial oil policy. But more important, Statoil's carried-interest agreements in all exploration and production operations enabled the company to grow into a giant in Norwegian economic terms. By 1980 the state oil company was recording pretax revenues of $28 million. About 31,000 b/d of equity crude were coming from its 44 percent share of the Statfjord oilfield alone. The company had expectations of controlling 35 million tonnes per year of equity and state royalty crude by the year 2000. This corporate success was partly owed to Statoil's 50 percent participation in five major oil and gas fields. Also, it was able to accept a real rate of return on investment of only 10 percent, while foreign oil companies required a 15 percent rate of return to offset the costs of unsuccessful exploration.[41]

Because Statoil was benefiting the Norwegian economy directly, the government was hard pressed to get rid of it. The company continued to pursue a strong policy of tying Norwegian company contracts to its production operations, thus directly stimulating Norwegian industry.[42]

The oil policy choices of the Norwegian government between 1968 and 1985 started with a surge of state-owned entrepreneurial oil activities. This was followed by increasing concessions to domestic industrial groups and growing reliance on revenues from international banks. But the state company remained financially strong. Although the Conservatives took power in 1981, they did not divest any major operations.

The initial period, 1968–75 was characterized by the government's assertion first of management control, through its administrative leasing policy, and then of revenue control, through corporate income taxes, royalties, and a special tax of 25 percent. But its assertion of ownership control through the creation of Statoil, the fully state-owned oil company, was the dominant policy of the period. This policy side-stepped earlier efforts by the government to participate in leases to make way for a full-blown entrepreneurial oil policy.

The period from 1975 to 1980 was characterized by policy concessions to private domestic groups previously excluded from oil wealth. The dominant government policy was managerial. The government was obliged to forego its ownership aspirations regarding oil subcontracting and tankers to assert management control over those activities then controlled by multinational corporations. It used informal preferences attached to lease agreements to concede a share of oil-related operations to Norwegian shipowners. The government then asserted some operational control over the multinationals' oil operations to grant adamant Norwegian fishermen a share of oil compensation for damages and to promise them strict safety regulations.

In the final period, 1980–85, the Norwegian government increased its reliance on external financing from international banks. It had to expand its oil activities as well as subsidize domestic industries harmed by oil-induced inflation and currency appreciation. During this period however, the Conservative government continued to pursue a strong entrepreneurial oil policy and used Statoil's managerial control to require the use of Norwegian industries and workers offshore.

This sequence of policy choices was strikingly similar to that in Britain. Across the North Sea, however, the responses of private industry were weaker than in Norway.

BRITISH OIL POLICIES

The British government's goal was similar to Norway's at the start—immediate oil exploration—but the two countries' domestic industries played different roles in bending policy outcomes toward their own interests. By the mid-1960s the British government was eager to gain energy supplies and revenues from its offshore oil and gas resources. It saw domestic oil production as a way to offset high oil imports and improve Britain's overall trade deficit (almost $1.8 billion in 1970). The government allowed multinational oil companies to develop oil but retained control of gas production. Gas supplies and prices were managed by the British Gas Council, the public utility operating the national gas system. The Labour government restricted gas exports because of domestic political pressure to tighten control over gas development. In 1968 and 1971 oil companies signed contracts for exclusive delivery of gas from southern fields in the British sector of the North Sea to the British Gas Council, giving a government-owned enterprise a monopoly over the distribution of British gas.[43]

With gas supplies firmly under government control, multinational oil companies focused on oil, where markets were free of government interference—distribution networks were controlled internationally rather than nationally. The oil companies did not begin producing oil in the British sector until 1971. After initial discoveries, in the late 1960s, the Conservative government leased offshore blocks to both national and foreign-based multinational oil companies to spread capital and technology risks. Through 1970 foreign companies, including Shell, Esso, and Occidental, licensed 62.5 percent of the area opened for bids. British Petroleum also gained a substantial share of blocks.[44]

The British government, unlike the Norwegian, refrained from intervening in the management of production rates. The government, anxious to gain immediate and substantial oil revenues, favored a rapid rate of extraction by the companies. It planned for production to reach 100–150 million tonnes by 1980, almost twice the Norwegian ceiling. This level promised to keep Britain a net oil exporter into the 1990s. Even a change in public opinion, followed by parliamentary legislation in 1974–75 granting production control to the new Labour government's Department of Energy, did not dissuade government policy from the encouragement of rapid extraction.

But the British government did join the Norwegians in choosing to administer licenses rather than auction them off. Administrative allocation enabled the Department of Energy in pursuing its oil policy to discriminate between foreign and national, public and private companies. Employment and service requirements could be attached to licenses. Furthermore, the decentralized "oil block" system, also used by Norway, enabled the British to minimize the time and increase the intensity of effort between commencement of exploration and ultimate relinquishment of a block.

The worsening state of the British economy, however, made government control of oil revenues a high priority. Before OPEC's initial 1971 price rise the government was cautious about fostering the interests of foreign oil companies in North Sea supplies. Taxing profits too heavily might reduce production advantages in the North Sea compared to other areas of the world. In the 1960s royalties on oil production remained at 12.5 percent. Area fee revenues were $15,000 per hundred-square-mile block over six years, increasing afterward until a ceiling of $174,000 per year was reached. Corporate income taxes were 45–50 percent, as in Norway, but transfer prices in Britain were set by the oil companies rather than being negotiated between the companies and government as in Norway.

But the opportunity for windfall oil revenues provided by OPEC price hikes and a worsening balance of payments brought the government to throw caution to the winds. In 1972–73 the Parliamentary Committee on Public Accounts called attention to the fact that foreign companies were reaping high profits by being taxed in Britain at one of the lowest rates in the world and then investing earnings in other countries. The government's stronger stance was partly a result of BP's exclusion from overseas production by OPEC politics and nationalizations. Taking the lead initiated by the Conservatives, the new Labour government in 1974 claimed a 70 percent tax share of oil revenues, similar to that of the Norwegian government. But the British also imposed a new Petroleum Revenue Tax (PRT) based on a flat rate of 45 percent for each field (deductible from corporate income taxes) rather than on residual profits after taxes and royalties as in Norway. By taxing all profits, this system prevented companies from offsetting expenditures against dues across fields. The following year the government also passed the Oil Taxation Act, which aimed to halt the drain on the British balance of payments caused by oil companies transferring tax loads from British operations to lower-taxed operations in other countries. By 1978 the Labour government had raised the level of taxation to 75 percent of North Sea profits. This resulted from a series of measures increasing the PRT from 45 to 60 percent and reducing production allowances and special capital allowances associated with that tax.[45]

British National Oil Company

It was also clear by 1974 that a radically new policy was needed to improve British government oil revenues and management of the domestic economy. Britain was experiencing the highest inflation rate, largest balance-of-payments deficit (1.8 billion, mostly due to oil imports), and fastest-rising unemployment rate since World War II. In response Labour adopted a new industrial policy, designed to use government management and financial incentives to generate a surge of new industrial investment and restructure the basic organization of British industry, including oil.[46] The Labour government believed reorganization of the North Sea oil industry could reduce inflation and decrease the trade deficit. In 1974 it called for a "fairer share" policy that would increase government revenues both by stricter taxation of the oil companies and by state participation in offshore operations.[47]

The previous sixty years had clearly demonstrated that the government could not impose its policy objectives on British Petroleum despite major state shareholdings in the company. BP functioned like any other privately owned multinational oil company. Any attempt by the government to interfere with corporate management or operations would have created serious problems with its private shareholders. In the fifth round of licensing, in 1974, therefore, the government first introduced the principle of 51 percent direct state participation in oil production. But this notion also required the government to assume 51 percent of exploration and development costs.[48]

In 1976 the government took a bolder step by establishing the 100 percent state-owned British National Oil Corporation (BNOC). The company was granted an option to take 51 percent of all future offshore licenses and to renegotiate past ones. By acquiring the offshore oil and gas interests of the National Coal Board and the Burmah Oil Company, BNOC gained an interest in or operating responsibility for sixteen offshore fields. By January 1978 the Department of Energy and BNOC had signed forty-two agreements with oil companies giving BNOC majority participation in fields licensed earlier.[49]

A major compromise between the state and private oil companies was apparent. Though BNOC could, in most cases, take up to 51 percent of the oil produced, the international oil companies could purchase it all back. And though offshore oil production by BNOC and multinational oil companies in 1977 reached 770,000 b/d, British government returns from the PRT, other oil taxes, and royalties remained low. Oil companies were allowed to deduct capital and other allowances before any taxes were levied on their production profits. As a result the government received only about $456 million in royalties and taxes from production in 1977, compared to the $2.02 billion projected by D. I. MacKay and G. A. MacKay in 1975.[50]

British Shipping's Response

Other British industries feared the expanding Labour government involvement in offshore markets. Traditional British shipowners, staunch supporters of noninterventionist Conservative policies, wanted to prevent government from interfering in shipping. Government oil operations might lead to such interference or to competitive state investment in oil-related shipping markets.

Traditional shipowners worried not only about state involvement but also about the dual role of international oil companies as both owners and consumers of British shipping services. After World War II multinational oil companies developed a dominant position in national shipping, both financially and in tanker capacity. By 1968, five out of the top eight British shipping companies were oil companies, Of those, BP, Shell, and Esso owned about two-thirds of the U.K.-registered tanker fleet, equivalent to two-fifths of total British tonnage between 1968 and 1973.[51] These vertical investments into British shipping gave international oil companies power over traditional British shipowners, because they made the multinationals both major consumers of and competitors for tanker services. Besides, traditional shipowners could not gain a foothold in the oil business because they depended on the oil companies' fluctuating and extremely competitive chartering. Foreign oil companies also would not give them long-term charters (more than five years) with provisions against rising costs.[52] Moreover, the oil companies' 40 percent share of British shipping gave them political influence within British shipping committees. As a result British shipowners, unlike the Norwegians, did not venture into offshore oil production or subcontracting.

While shipping and other domestic industries stayed away, the labor unions were pressuring the Labour government to engage British industry offshore. As Britain's economic slump worsened during the mid-1970s, unions and key national industries, such as construction, demanded that the government provide new employment opportunities. Vulnerable to pressures from these electoral constituencies, the Labour government—as its Norwegian counterpart had done—began applying minor pressure on the oil companies to use British companies for some subcontracting work.

In 1975 the secretary of state for Energy and UKOOA, the representative organization for the forty-two oil company operators, agreed on a memorandum of understanding and a code of practice. These "ensured that British industry is given a full and fair opportunity to compete for business in the UK offshore market." British industry was to provide a "major and progressively increasing share of the goods and services required for the development of the continental shelf. . . ."[53] The Offshore Supplies Office (OSO) was set up by the Department of Energy to implement these new constraints associated with leasing policies.

The government, by implementing such regulations through the OSO rather than using BNOC's own contracting discretion, signaled the low priority it assigned to stimulating private industry offshore. The Labour government was more concerned about its own oil production activities and had little reason to assist Conservative shipping with contracts, except in response to union pressure for offshore employment. The government had already spent its "political capital" with the multinational oil companies while negotiating BNOC participation offshore.[54] Although the OSO provided some incentives for traditional British shipowners to invest in drilling rigs and supply boats, BNOC made almost no effort to contract their offshore or tanker services. Thus the government's limited political resources and the mutual disinterest in cooperation between Labour's oil and Conservative shipping constrained the integration of national shipping into offshore oil.

Scottish Fishing's Response

As in Norway, domestic industries facing financial or operational constraints had to conform to state oil and gas policies in the North Sea to survive. Particularly after 1974 oil activities displaced Scottish fishermen. Between 1970 and 1974, 247 wells were drilled offshore, mostly within one hundred miles of the Scottish coast or the Shetlands. Where the two industries coexisted, oil installations and debris on the sea floor damaged fishing gear and five hundred–meter oil safety zones increased hazards to navigation. Oil pollution was also a concern. Up to 1974 opposition to oil by fishing interests was minor and had little effect on the Conservative government's policies. Government officials were estimating that revenues from oil production would be about seventy times those of fishing and therefore wanted to avoid conflicts with fishing that would impede oil development. In 1974 the Department of Energy set up the Fisheries and Offshore Oil Consultative Group (FOOCG) as a forum for fishing, oil, and government representatives to work out the political aspects of offshore oil displacing fishing.

But by 1975–76 fishing industry representatives claimed reductions in operating space of up to 20 percent. They estimated that between 1977 and 1978 the average loss of access would increase by another 25–30 percent. Though the economic losses were small by oil industry standards, the potential loss of access was significant enough to worry fishing industry leaders about the future of North Sea fishing.[55]

When Labour took power in 1974, however, fishing's electoral support became a source of industry leverage. The British Labour government's parliamentary majority was small, similar to the Norwegian situation, so the possibility that fishermen might shift their partisan support became politically salient. Of the twenty-two parliamentary seats associated with fishing constituencies in 1978, nine were held with majorities below 6 percent. Fishing constituencies thus controlled three times the average share of marginal seats, and Labour risked losing seven of them.[56] In addition, the rise of a Scottish nationalist movement since 1975 concerned the Labour party. Many Scottish fishing communities had Labour representatives, and a loss of parliamentary seats to the Scottish Nationalist party could weaken Labour party control. The potential partisan shift of fishing communities increased the industry's influence over the policies of the Labour government.

By 1976 Scottish and English fleets began organizing against the impending threat from oil and EEC fishery policies. The political reality was that the government was trading British fishing for oil and EEC concessions linked to foreign fishing in British waters. A strong fishing lobby in Parliament could pressure government regarding encroachment in these two offshore uses. The political reorganization that occurred followed regional lines and reflected divisions between the corporate and the share-owned segments of the fishing industry. In August 1976 the company-owned English

distant-water and Aberdeen fleets joined to form one unified
federation, the British Fishing Federation, representing corporate
fishing interests throughout Britain. Concurrently, Scottish in-
shore fishermen consolidated politically into the Scottish Fishermen's Federation to strengthen the influence of the Scottish
share-owned industry.[57]

Despite its increasing political power, fishing was not critical to
Labour government rule. Labour's parliamentary majority in Britain did not depend upon a small number of fishing votes, as it had
in Norway. Thus fishing was unable to stop oil; but it did succeed in
increasing compensation for oil damages and other fishing losses.
In 1976 UKOOA set up a compensation fund for fishing gear
damages, and as its administrating body FOOCG paid out $213,272
by 1978.[58] The oil industry made sure that no legal responsibility
was linked to compensation payment. Fishermen also had to prove
damages and provide evidence identifying companies responsible
for the damage, which resulted in fewer claims than in Norway and
thus less help to working fishermen. Scarce fish supplies nationally
in 1975–77 caused high prices, however, allowing British fishermen to offset temporarily the effect of low compensation.

Britoil Bails out the Treasury

The return of the Conservatives under Prime Minister Margaret
Thatcher in 1979 heralded a new era of monetarist Tory policy
aimed to reduce the Public Sector Borrowing Requirement, help
balance international payments, and stimulate domestic industry.
One such effort focused on raising oil and gas revenues through
increased taxation. Initially the Conservative government attempted merely to tighten the taxation policies introduced by Labour. It
raised the PRT from 60 to 70 percent, in response to the doubling of
international oil prices when Iranian supplies were cut back, and
increased the capital allowance to 35 percent. But by 1981 the
government was forced to introduce a new tax, the Supplementary
Petroleum Duty, in order dramatically to lower the Public Sector
Borrowing Requirement of the government. This tax was set at a 20
percent rate and was deductible from the PRT and corporate income taxes. The effect of the tax varied with the size and prof-

itability of fields, but it increased the overall marginal taxation rate to 90.3 percent from the previous 87.4 percent rate.[59]

The unintended consequence of stringent taxation in the five years up to 1982 was significantly to reduce oilfield development. Development in 1982 dropped by 13.9 percent compared to 1981; even so, production increased by 15.5 percent to 103.3 million tonnes, giving the government 21 percent more tax and royalty income, to a level of $22.5 billion. Oil companies argued that high production levels reflected fields developed under earlier, more favorable taxation regimes and that development would continue to decline. The Department of Energy forecast in 1982 confirmed this prediction. It projected a 1985 peak in total production of existing fields of 95–130 million tonnes, followed by a slowdown in development owing to U.K. taxation rates which would end British self-sufficiency by the late 1980s.[60]

These forecasts led to a major loosening of taxation policy by 1983, reducing the Treasury's share of oil and gas revenues. At first the government decreased its taxation rate from 70 to 60 percent, abolished the Supplementary Petroleum Duty, and replaced the latter with an Advanced PRT that increased the PRT from 70 to 75 percent. The overall effect of the changes was to reduce the marginal tax rate from 90.28 percent to 89.50 percent. By June 1983 the government had proposed a more radical slackening of taxation. It reduced the Advanced PRT from 20 to 15 percent and abolished royalties for fields approved for development after 1982. The government also granted large production, exploration, and pipeline allowances in association with the PRT. These taxation changes were expected to stimulate North Sea development into the 1980s by significantly improving the profitability of oil companies.[61]

The second major policy change instituted by the Thatcher government was "privatization." This policy involved both an industrial and a financial restructuring of British industry, including oil and gas. The government's reasons for privatizing were to reduce the Public Sector Borrowing Requirement and to shift industrial operations to the private sector to avoid political meddling in the public sector. The government also intended to trim the power of the trade unions within government-owned state monopolies and to increase competition within British industry.[62]

The impact on British state interests in oil and gas was substantial, but much less than what the Thatcher government had intended. The Conservatives spun off some of BNOC's activities into the private sector. But they could not afford to dismantle the company entirely, because its revenues were critical to the British economy. Unable to rely solely on tax and royalty revenues for government spending, the government was forced to abandon the idea of selling $900 million worth of BNOC oil by 1980 through a forward sale of oil. Instead the Department of Energy began to pump $1.2 billion in private capital into BNOC through sales of oil bonds, injection of risk capital, and the breakup of the company's trading and production operations. BNOC retained its ability to buy up to 51 percent of crude oil production for national security reasons (this fell to 49 percent after 1983), but its monopoly privileges were revoked.[63]

These changes aimed to convert the oil company into a mixed rather than a purely state-owned and single-operation rather than vertically integrated oil firm. By 1982, however, BNOC was the fifth-largest oil company in the world, with 1981 pretax profits of nearly $1.4 billion, which made it exceedingly difficult for the Minister of Energy to justify the sale of BNOC production operations to private interests.[64] The Department of Energy also tried to break up the monopoly of the British Gas Corporation (BGC, previously the British Gas Council) over gas distribution by allowing oil companies and consumers to negotiate deals independently. It also forced BGC to shift its oil interests to a new, temporary, state-holding company, Enterprise Oil, prior to offering them to private investors in hopes of raising an additional $1.3 billion for the British Treasury. But by 1980 the state companies, BNOC and British Gas, still controlled 70 percent of British-sector gas.[65]

In 1982 the Thatcher government made another valiant attempt to privatize BNOC. It created a new company, Britoil, separate from BNOC, which had enabled the government to buy 51 percent of all U.K.-produced oil. Furthermore, the government refused Britoil's request for a $579 million injection of government money into the company, which would have added to its $634 million in retained profits and $367 million from the government's National Oil Account. The requested money was designed to offset the company's liquidity problems and high debt-to-equity ratio, problems attributable to the government's request that surplus funds be channeled back into the Treasury through the same National Oil Account that subsequently granted the company its operating capital.[66]

Instead the government attempted to reduce the Public Sector Borrowing Requirement by selling off 51 percent of Britoil's North Sea oil and gas assets. The sale earned the Treasury about $1.6 billion. It also left Britoil an independent oil company with only minority rather than majority state shareholding. This loss of share actually gave the company greater fiscal freedom. As long as the government was the majority shareholder, any foreign loans to Britoil had increased the Public Sector Borrowing Requirement. But once state holdings dropped to only a minority, the company could assume foreign loans without their appearing as part of the Public Sector Borrowing Requirement.[67]

The company—both as BNOC and later as Britoil—had been very profitable since 1979. Net profits owing to equity-interest agreements grew from $101 million in 1979 to $339 million in 1980, falling slightly to $223 million in 1981. This profitability came largely because the company was not liable for the PRT or corporate tax paid by other private oil companies. It was therefore able to finance its 1980 development expenditures ($500 million) from equity—the proceeds from oil sales. By 1981 BNOC's equity production for its major five fields averaged 85,000 b/d. Participation oil was 618,000 b/d, third-party purchases 61,000 b/d, and government royalty oil 180,000 b/d. Total equity holdings in 1981 were 700 million barrels of recoverable oil in fields already producing or under development. By mid-1986 the government still had a major hold on oil and gas production through Britoil and British Gas. Despite a second sale of 243 million shares of Britoil, the government still retained a minority shareholding in the company. British Gas was still 100 percent state-owned, although the government was threatening to privatize it by the end of the year.[68]

The government was more successful in privatizing its holdings in British Petroleum than in Britoil. By 1986 sales of government shares in the company had reduced the state holding to only 31 percent from its earlier 51 percent and had raised $1.17 billion.[69]

By 1986 the government's total privatization of oil and gas had raised over \$2.4 billion for the British Treasury. This sum was equivalent to the amount the government had hoped to raise *each year* between 1984 and 1988 in a combined fiscal and industrial strategy (the capital raised would be used for current spending). That \$2.4 billion was less than the over \$6 billion already earned by the Treasury through sales of shares in nonoil public corporations. In many cases, such as BNOC, BP, and British Telecom, state companies were extremely valuable and profitable ventures (the government's 51 percent share in British Telecom alone was worth \$11.6 billion). The nearly \$5.8 billion in sales of publicly owned assets between 1979 and 1983 reduced the public-sector deficit by one-third. As of January 1984 the government planned to privatize in telecommunications, airways, airports, nuclear engineering, naval shipyards, and shipbuilding—but *not* Britoil or British Gas. By June 1986 Britoil and British Gas were back on the list.[70]

All in all, these privatizations indicated the Treasury's increasing grasp on British government policy and, in particular, on nationalized industries. Specifically the Treasury attempted to change laws for nationalized industries. Pre-1984 controls already gave the Treasury the ability to limit external financing for state companies and to require certain levels of return on assets per year per corporation. New controls, however, might enable secretaries of state to dismiss heads of state companies and ministers to order privatization of any company (current statutes require some companies to be kept intact). Furthermore, finance ministers might gain the right to insist on standarized accounting rules designed to stop profit and loss dodges practiced generally by both public and private companies.[71]

In sum, the British government's oil policy choices were strongly entrepreneurial by the mid-1970s. They became increasingly concessionary to private domestic industrial and financial interests once the difficulties of the British economy began to erode oil gains. This concessionary period resulted in fewer industrial policy changes than in Norway, but it did lead to a much stronger effort to privatize oil.

The initial period, 1968–76, was characterized by less interventionist management and revenue control by government. The Department of Energy did engage in administrative licensing but did not set a production ceiling as Norway had. The PRT was added to initial royalty, area-fee, and corporate income tax revenues in order to boost the British Treasury's oil income. The government did, however, create a fully state-owned oil company, BNOC, in order to supplement tax revenues with a direct source of government profit. BNOC participated through carried-interest agreements that gave the company a 51 percent claim to the oil produced.

The 1975–80 period in Britain involved fewer concessionary policy changes than in Norway. Labor union pressure forced the Department of Energy to pressure foreign oil companies to employ British industry and labor. Scottish fishermen also compelled the government to provide some oil compensation for damaged fishing gear in return for electoral support of the governing Labour party. Although policy changes were similar to those in Norway, therefore, their enforcement was much less stringent.

The dominant policy change that characterized the 1980–85 period was the Conservative government's effort to privatize all oil and gas operations. But this effort was, regarding BNOC, only partly successful. The government succeeded in selling off about $1 billion of the state oil company's trading operations and converting the company into a new minority state shareholding company, Britoil. But the government did not succeed in getting rid of the company entirely. Instead Britoil's profitability and its rank as one of the largest oil companies in the world made it too valuable a source of income and a holding of oil and gas assets for the Treasury to divest completely. The government did, however, reduce its public shareholding in BP to 31 percent. It also shifted the oil assets of British Gas over to a new state holding company, ready to be sold to the private sector.

The government had attempted to tighten taxation policy to gain more revenues for the Treasury, but by 1984 it had to slacken its grip. The Supplementary Petroleum Duty and Advanced PRT were aimed at and succeeded in increasing government revenues. But they also caused a slowdown in development activities by foreign multinationals. These activities picked up in 1984 only when the government significantly reduced the taxation rate to favor exploration and development into the 1990s.

INDONESIAN OIL POLICIES

The case of Indonesia, in contrast to Norway and Britain, shows the persistence of a government pursuing extensive state entrepreneurial investments in oil. Domestic private-sector interests in oil were weak, but strong multinational corporate interests, both Major and Independent, provided a mixed blessing for state oil enterprise. The result was a reliance on sovereign entrepreneurial policies even greater than in either Norway or Britain.

After national independence in 1949 the Indonesian government continued the contracts of three mutltinational oil companies originally brought in by the Dutch (Royal Dutch Shell Group; Stanvac, equally owned by Standard Oil of New Jersey and Mobil; and Caltex, equally owned by Standard Oil of California and Texaco). Despite President Sukarno's interest in reducing the foreign dependence of Indonesia, and despite general policies prohibiting foreign investments in other sectors, oil production was an important source of tax revenues. No national capacity existed to replace foreign companies immediately.

Although foreign oil companies had been working in Indonesia since 1885, the Indonesian government granted them a new type of service contract, called a work contract, in 1963. Under these new contracts, profits were to be divided 60–40 in favor of Permina (later Pertamina), the government's oil company. In fifteen years the foreign oil company was to be fully operated by Indonesians. However, the government could neither exercise management control nor claim more than 20 percent of its profit share in oil supplies.

It was also during the 1960s that the Indonesian government placed all oil activities under the control and operation of the state. Law 44 of 1961 stated that all mining of oil and gas would be undertaken by the Indonesian government and carried out by state enterprises (Art. 3). "Mining undertaking" was defined in Article 4 to include all activities of an integrated oil company: exploration, production, refining, transportation, and marketing. Legally, only a state oil company such as Permina (later Pertamina) could run operations related to Indonesian oil production. To operate in the country, foreign companies had to contract from the state company. Although the policy was carried out less stringently than

worded, it did set the tone for state control of Indonesian oil operations. The later passage of Law 8 of 1971 threatened to reverse this
policy by bringing under the control of the president state oil company investments in sectors not directly related to oil (e.g., shipping). Furthermore, the new law created a supervisory financial
council to oversee investments by the state oil enterprise.[72]

State enterprise was thus integral to the oil policies of the Indonesian government from the start. The government had seen that
state oil production would secure a domestic source of oil as well as
earn foreign exchange through exports.[73] It was also clear that
state enterprise would increase the national share by providing the
government with direct profits and control and by expanding the
productive capacity of Indonesian industry.

Pertamina

Between 1957 and 1961 the Indonesian government formed
three national oil companies in order to create a nationally owned
share of oil and gas production. By the late 1960s, however, it
became clear that the government could control a single oil company more effectively than three. Hopes were that one company
could expand the national share to more than its 10 percent of total
production and, in particular, increase domestic supplies and earn
foreign exchange through oil exports. In 1968 one of the national
companies was eliminated and the other two consolidated into a
single, 100 percent state-owned oil company named Pertamina.
Company leadership was chosen from the Indonesian Army, which
also benefited indirectly from the company's finances.

The oil Majors controlled most onshore production, through
Caltex, Stanvac, and Royal Dutch Shell operations; however, offshore oil and gas resources offered a new avenue for the pursuit of
state interests. In 1960 the government had claimed the archi-
pelagic waters surrounding Indonesia (with boundaries set along
straight lines connecting the outermost points of the most pe-
ripheral islands). This claim had established the initial basis for
territorial sovereignty offshore. The area enclosed 666,000 square
nautical miles of international water. Indonesian Law 11 of 1967
stated that the entire archipelago *and* its continental shelf were

specifically under Indonesia's mining jursidiction. To avoid confrontations with neighboring countries regarding its claims, the Indonesian government signed offshore boundary delimitation treaties with Malaysia, Thailand, and Singapore, giving the government undisputed sovereignty over its offshore resources. Treaties with Malaysia and Singapore in 1969 and 1971 delimited the exact extent of Indonesia's sovereign claim to waters in the Strait of Malacca and the South China Sea.[74]

This extended sovereignty over oil resources gave the state access to offshore production without jeopardizing the onshore investments of the Majors. It also gave the Independents access to Indonesian offshore resources without having to compete directly with the Majors, because Permina created new production-sharing contracts with the Independents in the late 1960s. The company thus gained the capital and expertise it needed to develop nationally owned oil operations offshore. The agreements were less favorable to foreign oil companies than were the work contracts held by subsidiaries of Majors such as Caltex. Foreign oil company contractors were allowed to subtract a maximum of 40 percent in costs (cost ceiling) from total gross production before production shares were divided up. The remaining 60 percent was then divided 65–35 in favor of the Indonesian government. By 1968 this arrangement had given Pertamina, the government's new company, control over 39.6 percent of offshore oil. However, a clause in all production-sharing contracts stated that Pertamina was also entitled to any oil produced at lower costs (up to 40 percent of the total), but some revenue from sales had to be returned to the contractors. Pertamina could control, sell, and therefore also transport up to 80 percent (39.6 percent + 40 percent) of total Indonesian crude oil production, but it had to return part of the 40 percent "cost oil" revenues to oil company contractors.[75]

Production sharing had the strong support of the Indonesian government, the Independents, and the Japanese government. The Indonesian president and the army agreed with Pertamina's head, Ibnu Sutowo, that production sharing offshore would increase Indonesian ownership, management, and operational control within the country's oil industry. For Independents such as IIAPCO, Japex, and AGIP, access to high-value Indonesian crude production through production sharing was the only way to break the Majors' monopoly over Indonesian operations. American and Canadian Independents were interested in expanding into overseas production to gain increased oil supplies for U.S. and Canadian refineries. These companies knew that although operating costs were higher offshore, offshore production also meant less government interference, fewer labor disruptions, and easier transportation than onshore.[76]

For Japan, production-sharing agreements with Japanese Independents such as Japex meant a much-desired foothold in Indonesian oil development. This bilateral cooperation was strongly supported by the Japanese government. Japan's Foreign Office was particularly interested after 1968 in reducing Japan's energy dependence by expanding Japanese investments in energy resources overseas and loosening the Majors' control over 80 percent of oil marketing within Japan.[77]

Production-sharing contracts thus gave Pertamina and the Independents an opportunity to create a new oil market offshore, one that unlike the onshore market was not controlled by the Majors. Between 1966 and 1971, forty-two production-sharing contracts were signed with Japanese, Canadian, American, and other Independents. These contracts put more than 90 percent of Indonesia's offshore waters under oil exploration or production.[78]

But the Indonesian state was not satisfied with a share of oil production. The government also gave Pertamina approval to expand into a fully integrated state oil company with national, but not yet international, shipping and foreign marketing operations. The company intended to gain the benefits of vertical integration, reducing capital and operating costs and securing an oil supply to Japanese markets. Law 44 legitimized such state control of integrated oil operations.[79] In 1966 Permina (later Pertamina) and Japanese business and government interests formed a joint venture with 50:50 ownership, the Far East Oil Trading Company, to market Permina's own crude and its 20 percent share of Caltex's crude in Japan.[80] By 1968 Pertamina had created a tanker subsidiary, Ocean Petrol Ltd., in Hong Kong, and by 1970 the state oil company was exporting 46 million barrels, about one-sixth of Indonesian oil production in that year (311 million barrels), to Japanese markets. In addition, by 1972 Pertamina had built up a major tanker fleet of about fifty-five tankers, and its subsidiary, Pertamina Tongkang, owned a fleet of one hundred supply boats.[81] This cross-investment of the state oil company in shipping was not opposed by Indonesian shipping, also government-run. The government planned to build the Indonesian tanker fleet through the state oil company rather than through direct government investments.[82]

By building up its national fleet of tankers and supply boats and by forming a joint venture for marketing in Japan, Pertamina was by 1973 almost a fully integrated international oil company. All it lacked was international tanker operations. The benefits of integration were reduced capital and operating costs and a secure oil supply line to Japanese markets. Independent oil companies, American banks (including Chase Manhattan and Citibank), and even the Japanese government hoped to share in these benefits by providing Pertamina with the capital and expertise it needed.

Opposition from Major Oil Companies, International Banks, and Government Economists

However, the attempt by the state as entrepreneur to control the Indonesian oil industry ran counter to the interests of the Majors, international banks, and the U.S. government. The Majors wanted to retain their profits and exclusive control over integrated production, international transportation, and marketing operations linking Indonesia to Japan and the United States. Pertamina's effort to share this control was a threat. The Majors were alarmed that if Pertamina were to expand internationally, the company might erode their dominance over the Indonesian oil trade and in particular their role as exclusive suppliers of high-grade Indonesian crude.[83] International banks and loan agencies such as the International Monetary Fund were also worried. The Indonesian state could avoid international rules of accounting if it succeeded in using its reserves of oil and gas offshore as collateral for Pertamina's oil investments.[84] Finally, confronted with the 1973 oil crisis, the U.S. government wanted Indonesian oil supplies going to the United States and Japan to remain primarily in the hands of the major multinational oil companies.[85] This coalition of international groups was able to pressure the Indonesian government regarding Pertamina's financial situation, which the government guaranteed. One American bank recalled a short-term loan that Pertamina could not pay off, starting a wave of loan payment calls by other banks. The company came near bankruptcy because it was already shouldering nearly $10 billion in debt.

But elites within the Indonesian state itself also opposed Pertamina's oil policies. The economists in the Indonesian ministries of Finance, Mines, and Planning (BAPPENAS) were particularly concerned. Pertamina's expansion would increase the company's hold over them financially by allowing it to reinvest oil funds. By 1973–74 taxes from foreign oil companies were providing 30 percent of all government revenue. Despite 1971 legislation (Law 8) increasing government control over taxes on foreign oil companies, Pertamina was still able to retain most of those funds for reinvestment. In addition the government economists had little influence over Pertamina's policies or use of funds, and they attributed the company's large debt to mismanagement and fraud. However, the company was also powerful within the Indonesian economy, so powerful that by 1974 the foreign press was calling it a state within a state. Finally, Pertamina's investments and payments throughout Indonesia were altering the effect of the government's regional economic policies.[86]

A conflict of significant international dimensions emerged over who would ship Indonesian oil. In 1973 Pertamina ordered fourteen ocean-going tankers to transport its 800,000 b/d oil supply (its own crude and its 65 percent stake in production-sharing arrangements).[87] This was the initial step in integrating the company's tanker operations at the international level. The Majors had thus far held exclusive ownership and management control over Indonesia's international oil transport, and the step provided foreign interests with an opportunity to topple the company. In 1974 foreign banks called for debt payments on the short-term loans keeping the company afloat. Pertamina defaulted, and the threat of bankruptcy loomed.

The Indonesian government was forced to make major concessions in its offshore oil policies, particularly regarding state enterprise. The main concessions involved an increase in production-sharing contracts but a decrease in state entrepreneurial activities, The Ministry of Finance took over the state oil company and re-negotiated all production-sharing contracts with foreign oil companies to increase the state's share from 65 percent to 85 percent.[88] Production-sharing contracts retained this 85–15 split into the 1980s.[89] At the same time, however, the government reduced the company's production level by halting all new exploration and transferring Pertamina's onshore production to foreign oil companies. The new head of Pertamina (previously the minister of finance) then divested the company of international transport operations so as not to interfere with the interests of the Majors in those areas. Pertamina retained half of its national tanker and all of its subcontracting (supply-boat) operations. By 1982 the state company still owned forty-six oil vessels.[90]

The Japanese, who had been supporters of Pertamina, did particularly well for having been on the losing side. The government gave Inpex, the main Japanese Independent, responsibility for production on Pertamina's onshore fields, making Inpex-Total production the country's highest at 1.3 million b/d.[91] Despite this gain in Independent production, the Majors retained previous production levels and their internationally integrated transport operations.

As a result Indonesia's entrepreneurial policies were curtailed but not cut back to the production stage (which was as far as Norwegian policy had reached). Indonesia still had an ambitious production-sharing policy through which to channel its leasing, taxation, regulation, state investment, and entrepreneurial endeavors. Pertamina still acted as a vertically integrated company, its operations including production, transportation, subcontracting, and marketing. Except for overseas marketing, however, these operations were primarily national.

In sum, the Indonesian state's interests in expanding its economic control and industrial capacity through state enterprise were diametrically opposed to the interests of the Majors. State interests did coincide with those of the Independents and Japanese investors, but these were not the source of the state's entrepreneurial interests, merely catalysts providing capital and expertise needed for the pursuit of those interests. The state split between the interests of the president, military, and state oil company, on the one hand, and the interests of the Ministry of Finance, on the other. The former supported the pursuit of state interests through entrepreneurial policies, while the latter preferred taxation and regulatory policies. The interests of the Ministry of Finance may have coincided with the interests of international oil and finance, but the interests pursued by the president and the state oil company were clearly nationalistic, seeking oil industrial development through expanded state enterprise.

Indonesian fishing was quite a different story. The industry lacked both the industrial linkages and finances that would allow it to integrate horizontally into oil and the political clout in the Indonesian Parliament or government bureaucracy needed to demand compensatory benefits; it was displaced offshore. By 1970–71 more than 90 percent of offshore Indonesia was under exploration or in production. The most intensive oil activity was in the Java Sea, the Straits of Makassar, and the South China Sea off Kalimantan. In response to oil company requests for six hundred-square-mile exclusive oil activity zones, the government prohibited small-scale fishermen from certain traditional fishing areas. More intensive Japanese-Indonesian joint-venture trawler companies were relegated primarily to eastern, nonoil areas—a regional rather than a local displacement.[92]

Central Financial Control in the Indonesian State

By 1980 Indonesia was still in a very strong economic position regarding oil. Rapid price increases in 1979 for crude and oil products had strengthened the government's fiscal budget and provided foreign exchange reserves at a record high of $4.14 billion. Even if crude sales were to fall, the government expected that petroleum earnings would rise 92 percent by the end of 1980 because of increasing prices, new liquefied natural gas marketing, increased domestic demand, and improved production-sharing terms. This increase would expand the oil and gas portion of the government's income from 48.2 percent in 1979–80 to 60.9 percent in 1980.[93]

The Majors had meanwhile regained the lead in Indonesian oil production. Pertamina had seventy-five contracts with foreign companies, and the leading foreign contractors were Caltex, Stanvac, Calasiafic, and Topco—a shift away from the dominance that Japanese Independents had been able to assert in the late 1970s. The leadership of the Majors was demonstrated by the fact that the value of foreign company oil exports under work contracts was higher than those for production-sharing contracts: $5.52 billion compared to $3.87 billion.[94]

The years of oil glut, 1980–84, began to strain the government's oil and gas policies. OPEC mandated that its members, including Indonesia, decrease their production levels in order to sustain high prices despite oversupply on the world market. Indonesia was required to drop production from its 1.6 million b/d in 1981 to 1.3 million b/d in 1982. Diminished production reduced the government's corporate tax revenues from foreign producers, revenues which had increased on average by more than 50 percent per year between 1979 and 1981; they constituted 70 percent of domestic oil revenues in 1981. But by 1982 the effect of production cuts had already begun to decrease corporate tax income, to 66 percent. These corporate tax decreases strained the government's policy of using cheap production-sharing crude for domestic use. The government was forced in 1982–83 to cut its subsidization of low prices for domestic crude. Rising local demand for oil had to be rationalized by higher prices, which necessitated a 60–75 percent domestic fuel price rise to offset the losses in corporate taxes. By 1984, however, 73 percent of foreign exchange was still being earned from oil and gas exports.[95]

Despite an oil glut on the global market, the Indonesian government chose to stimulate increased exploration for new oil and gas fields. The government was intent upon sustaining production levels in the face of rapid depletion of its oil fields. It demanded that companies spend a minimum total of $327.3 million over the next six to ten years. Companies were asked to pay $1–15 million in information bonuses and $2–19 million in production bonuses on oilfield discoveries, rising to $15 million per field at 100,000 b/d.[96]

But the government was suffering from other constraints on its oil income. Half of Indonesia's total crude production was still controlled by Caltex. The company was producing that oil from the onshore Minas field and selling it under a two-tier price system discriminating by 50 cents per barrel against Japanese importers. Purchases were made through Pertamina's Far East Oil Trading Company, Japan Indonesia Oil Company (50 percent Pertamina-owned), and Perta Oil, as well as direct sales by Caltex. Discriminatory pricing of Caltex's crude was forcing Japanese companies to cut back on their spot purchases of Indonesian crude and replace them with politically uncertain Middle Eastern imports. Long-term contracts with Japanese customers, however, were continued. This discriminatory pricing was reducing total government income from oil sales.[97]

Furthermore, Caltex's control of the Minas oilfields directly affected the government's tax revenues from oil. Minas was used as the official marker crude for oil prices and thus was indirectly the marker for acceptable taxation rates. But OPEC also had direct influence over Indonesia's price. Without OPEC consent, prices on Minas crude could not be lowered, and relative tax revenues thus increased because of the relative gains from greater sales.[98]

With its hands tied by Caltex and OPEC, the Indonesian government became increasingly dissatisfied with the terms of its old work contracts, under which Caltex operated. Pertamina began to renegotiate with Caltex, insisting that the company shoulder most of the burden of OPEC's mandated reduction in Indonesian production. The government demanded that the company drop its production from 850,000 b/d—the level Caltex had sustained throughout the 1970s—to 650,000 b/d. The new eighteen-year agreement replaced a twenty-year work contract that expired in 1983. It stipulated that the after-tax profit split would be 88 percent for Pertamina and 12 percent for Caltex. This was a slight improvement for Pertamina over the typical 85–15 split in production-sharing contracts. Over the new contract period Caltex agreed to invest $3.06 billion in new exploration and enhanced production.[99]

Another government effort to increase oil income was to reshuffle Pertamina's leadership in 1984. President Suharto's changes seemed to suggest a shift away from the restrictive marketing policies under the former director of the budget in the Ministry of Finance (Haryono) and under his successor, Sumbono, which were deterring Japanese consumption with discriminatory oil pricing. The new head of Pertamina was Ramly, the former head of the state-owned tin company. Suharto hoped that he would revive exploration and improve marketing so that Indonesian excess refining capacity could be filled with domestic and foreign supplies. In addition, the former head of Pertamina's liquefied natural gas project was assigned to be director-general of the Ministry of Mines and Energy, displacing an incumbent of six years. The minister of mines and energy in 1984 directed Pertamina to serve the government's oil revenue and development efforts more closely by maximizing its foreign exchange earnings through exports and lowering production costs and domestic fuel subsidies. The company was also instructed to channel oil revenues to the central government to finance development plans and to submit its records to audits by the government and by such outside auditors as the IMF.[100]

The Indonesian government's entrepreneurial oil policies between 1968 and 1985 were the most aggressive of our four countries. Indonesia did make major concessions to international banks and the Majors, however, particularly during the mid-1970s. These concessions sustained investments by the state oil enterprise at the national level but curtailed them at the international level.

Between 1968 and 1976 the government formed Pertamina and asserted wide ownership, management, and revenue control over Indonesian oil and gas operations. Within the first four years the state company reached a production level of 140,000 b/d. It was also fully integrated into subcontracting and transportation at the national level and ocean-going tanker transport and marketing at the international level. Indonesia thus differed from Norway and Malaysia, where state enterprise extended only to national oil and gas production. Furthermore, Law 44 had established that the state would undertake all petroleum operations, with foreign companies hired as contractors. This law plus the introduction of production-sharing contracts gave the Indonesian state, through Pertamina, effective management control over leasing and levels of production. Work contracts and production-sharing contracts also gave Indonesia revenue control over nearly 80 percent of Indonesian oil production.

But the expansion of Pertamina, which soon made it one of *Fortune*'s five hundred largest corporations, came to an abrupt halt in 1975. The 1975–80 period was characterized by a sharp decline in state ownership control but a surge in revenue control to offset financial losses. The coalition of Indonesian economists in the ministries of Finance and Planning (BAPPENAS), backed by IMF and U.S. government pressure and spurred on by the threat of Pertamina bankruptcy, convinced the president and the army to trim back on Pertamina operations and reorganize the company's leadership. The new leadership divested international tanker operations and cut national shipping in half. The government then increased its share of oil revenues through improved production-sharing terms (60 percent up to 85 percent) to offset Pertamina's income losses. However, the state's production share went directly to the Ministry of Finance rather than to Pertamina.

This tightened grasp of the central government over Pertamina continued into the 1980–85 period. The government continued to increase its revenue and management control over oil operations. Improved production-sharing terms (from 85 to 88 percent) and required information and production bonuses increased government earnings by 92 percent. But production cuts required by OPEC after 1982 began to erode corporate income tax revenues. The government managed to drive a bargain with Caltex in which the state gained an 88 percent share of oil revenues and Caltex agreed to cut its production level in order to offset Indonesian income losses from OPEC cuts. The state's increased management control over production levels was reinforced by a 1984 change of leadership in Pertamina and at the Ministry of Mines, indicating that state intervention in oil operations might again pick up during the mid-1980s.

MALAYSIAN OIL POLICIES

The Malaysian state was ultimately more successful than the Norwegian, British, and Indonesian states in achieving ownership control of state oil and gas operations, both national and international. It resembled the other three governments in its eagerness to develop offshore oil as a source of revenue and energy supplies for national industrial development.[101] Shell was the only company producing in Malaysia up to 1963, while the country was still a British colony. Even after independence the government retained the company's concession contracts, because there was no national industrial capacity or expertise upon which to draw. The government also needed foreign capital and risk taking for oil development. By 1970 annual crude oil production in Malaysia was 6.56 million barrels.

But the Malaysian government had little control over the development of its own oil and gas resources. Shell's operations were in Sarawak and Sabah. These two states on the island of Borneo were officially under central government control after 1963, but they acted independently in the development and management of offshore resources. Sabah, for instance, had its own oil company, Sabah Energy. The lack of central government control led the Malaysian government to grant licenses to Esso, Continental Oil, and Mobil to explore off peninsular Malaysia, which was under the control of the Kuala Lumpur government. However, only Esso retained its concession, making discoveries in 1974.[102]

The discovery of oil and gas off peninsular Malaysia presented resources that the government could control free of the opposition of strong regional governments. At the same time the world oil crisis of 1973 demonstrated the importance of a secure domestic supply of energy. The government foresaw other advantages in controlling its own source of oil and gas. Domestically produced oil might reduce Malaysia's vulnerability to the international price-supply wars being waged among world oil producers and consumers. Furthermore, the government saw advantages in maximizing government oil revenues by generating state oil profits in addition to bargaining with foreign oil companies for higher taxes or royalties. State oil company profits could supplement these taxes and royalties and provide indirect financial spin-offs for other domestic industries developing in Malaysia. State oil investments could also create business opportunities for other state enterprises, such as the Malaysian International Shipping Company (MISC, 61 percent government-owned).[103]

Petronas

In 1974, the government formed Petronas, a 100 percent state-owned oil company, and began exploration in peninsular Malaysia where Esso had just discovered natural gas. Although the prime minister's Economic Planning Unit and Petronas favored cooperation with the Majors, they also wanted to secure a share of operations and control for the state oil company. The government, therefore, rescinded all oil concession leases to Shell and Esso and started renegotiating oil company contracts to give Petronas 59 percent participation through production-sharing contracts similar to those in Indonesia. It also set up the Production Development Act of 1974, which carved out a share of management control for Petronas in production as well as refining and transportation operations, over which foreign companies had previously had exclusive control.[104]

Resistance of the Majors to Malaysian LNG Transport

Having just formed Petronas, the prime minister's Economic Planning Unit pressured MISC into investing in five liquefied natural gas tankers, totaling 340,000 tons, to carry Malaysian gas to Japan.[105] This investment implied a dual government strategy for the development of Malaysian shipping. Investments in gas-carrying tankers were not likely to jeopardize the interests of the Majors in retaining control over international oil transportation. Brunei, the independent state between Sarawak and Sabah, already owned seven LNG tankers and had had no trouble from foreign oil companies. The Malaysian government intended to increase its foreign exchange earnings by exporting gas. It could also retain income by transporting the gas in its own LNG tankers rather than paying for foreign shipping.[106]

The other part of the government's strategy for MISC investments was a diversion from traditional shipping. Most less developed countries wanted to penetrate the barrier to their shipping posed by the developed country liner conferences that dominated international trade. Voicing their demands thorugh the UN Conference on Trade and Development (UNCTAD), less developed countries insisted that they carry at least 40 percent of their trade in their own ships. Although MISC and the Ministry of Trade supported this effort, the Malaysian government did not officially endorse the UNCTAD effort. Instead, it diverted the orientation of MISC's strategy for expanding its share of cargo transport from a quantity to a goods basis. MISC would expand its operations not by fighting for a share of liner trade but by increasing its transport of LNG gas, which was more profitable anyway.[107]

In other words, MISC could develop its fleet tonnage faster through the oil industry trade than through the international dry cargo trade, which was controlled by foreign liner conferences. LNG transport would also provide the national shipping company with high profits and give the government vertically integrated investments through Petronas's production and MISC's transportation. The Ministry of Trade and some interests in MISC itself were reluctant to divert Malaysian shipping further into oil operations, however, and the LNG tanker investments angered them.[108]

The LNG tankers were to service an LNG refining plant being built off Sarawak by Shell, with Japanese and Asian Development Bank financing, at a cost of U.S. $3 billion. Use of MISC tankers provided Japanese financiers (who were also backing MISC) as well as the Malaysian government with an opportunity to link their equity shares in refining, transportation, and production. However, some acceptable arrangement concerning MISC transportation of gas had to be worked out with Shell, which operated the refining plant.[109]

A major conflict arose over the extent of state oil enterprise. Though willing to concede a share of production control to Petronas through production-sharing arrangement, Shell was unwilling to share equity, the management of transportation, and the refining of LNG or oil. Earlier, in 1968, the foreign company had thwarted an effort by MISC to transport oil by refusing to contract its tanker.[110]

Between 1974 and 1977 deliberations continued between the Majors, Petronas, and the government concerning the Shell Bintulu LNG refinery and the transportation of its gas to Japan. The government came under increasing financial pressure to find contracts for the LNG tankers in which MISC had invested, backed by

Malaysian government guarantees on loans from the French government. The total cost of the tankers was U.S. $750 million, particularly sizable when one considers that MISC's total investments up to 1974 came to only U.S. $166 million. Delays in the negotiations already meant that LNG would not be ready for shipment until 1983, which would create a three-year lapse between tanker delivery and use; the idle period would cost MISC about U.S. $180 million.[111]

The severe financial consequences of these delays forced the government to give up a share of management control over refining and transportation in exchange for an equity share (of 65 percent) in the former and exclusive contracts for MISC in LNG transport. At the same time the government also repealed the national management shares requirement of the 1974 Petroleum Development Act. The act had originally established the legal basis for state control in the oil industry, and the repeal exempted refining and marketing from production-sharing licenses. Furthermore, the government agreed to keep Malaysia only a marginal exporter of oil, thus disqualifying it from OPEC membership.[112]

Despite its loss of exclusive management control, Petronas did retain legal ownership of all products and a substantial share of oil and gas supplies. Foreign oil companies could take a maximum of 20 percent in production costs recovery. Then shares were divided up into 54 percent (5 percent royalty plus 49 percent production share) for Petronas, 5 percent royalty for the government, and 41 percent (20 percent costs recovery plus 21 percent production share) for the foreign oil operator. Production-sharing contracts for twenty-year terms were signed in 1976 with both Shell and Esso.[113]

By 1982 the Malaysian government had through Petronas gained equity participation and a management share, but not exclusive control, in both Shell's and Esso's oil and gas operations. Although Malaysia's official production ceiling remained at 400,000 b/d, actual production stayed below 300,000 b/d in 1981, keeping Malaysia too small an exporter to join OPEC. Oil represented only 10 percent of Malaysia's exports.[114]

By 1978 MISC had become the sole contractor for LNG tankers

servicing the Sarawak LNG plant, jointly owned by Petronas (5 percent), Shell Group (17.5 percent), and Mitsubishi (17.5 percent). Five LNG tankers were in operation by 1983, when LNG production commenced and exports started flowing to Japan under a twenty-year contract.[115]

Shell and Esso continued to be the only Majors in Malaysia. They retained almost total control over ocean-going oil tanker operations. By 1978 Esso had just started production on its two offshore fields in peninsular Malaysia. Production was 60,000 b/d, but the estimated potential of the fields was 350,000 b/d; Esso was exporting its share. But consistent with the new production-sharing agreements, the Malaysian government's share was being consumed domestically rather than exported. Overall, Esso and Shell exported about 60 percent of their supply to Japan. Exxon and Mobil still ran the tankers used for export, while MISC's one tanker continued to ply the Middle East trade.[116]

In domestic waters the Majors also continued to have almost exclusive control over such secondary operations as supply boats and drilling rigs. Although the government had exerted some pressure to use national services, this was mostly symbolic. Foreign companies were requested to incorporate locally but not pay taxes. They could continue to use their own technology, experts, and capital. Of the fifty-four foreign subcontracting companies providing drilling and supply boat services in August 1978, three-quarters were American and not one was Malaysian-owned. The total value of subcontracts for Esso alone in 1977 was U.S. $21.6 million for rigs and U.S. $14.2 million for supply boats. So foreign companies also retained control of servicing operations for production.[117]

To sum up, by 1978 government oil interests (Petronas and the prime minister's Economic Planning Unit) and the Majors were in control of offshore production, and mixed government-private shipping (MISC) had horizontal investments in international LNG transport. But the Majors retained management control over international transportation of both gas and oil. They were willing to give up a share of production control but not of control over transportation and refining. The government finally accepted an ownership and management share just in production, in exchange for an equity share in refining and exclusive contracts for MISC in transportation.

Indirect Oil Benefits for Malaysian Fishing

For reasons similar to those in the other three cases, Malaysian fishing had to resist displacement by oil. Nor were domestic politics particularly favorable to fishing. Malaysian fishermen had some political role, which they owed to government redistribution policies between 1971 and 1980 that gave Bumiputra ("Native") Malaysians proportionately more economic control in the national economy.[118] But Bumiputra fishermen still held an insignificant position. Conversely, both Chinese-Malaysian and Japanese fishing companies had the capital and the technical know-how, but they lacked the political clout to influence the policies of the central government. As a result trawler fishermen, in particular, were simply displaced as oil drilling areas came to cover about 36,500 square miles. Shell's and Esso's operations, respectively forty-five and one hundred fifty miles from shore, primarily affected Bumiputra trap fishermen on the east coast of peninsular Malaysia and Japanese-financed prawn trawling vessels off the northern Borneo coast. During 1976 Esso's seismic surveying and exploration drilling damaged about nine hundred traps off the east coast of peninsular Malaysia, representing a loss of 135,000 kilograms per year of fish, valued at U.S. $159,000 per year. Pipelines and other oil industry structures also left large areas of the Borneo coast inaccessible to trawlers.[119]

Lacking political organizations or lobbies to demand payment for damaged gear and losses of catch, Malaysian fishermen had to depend upon other politically influential groups to act on their behalf. Specifically, Bumiputra political groups favored greater economic benefits for native Malaysians, of which fishermen were a large and poor group. Bumiputras were powerful in government but not in the economy, which was controlled mostly by foreign and Chinese-Malaysian interests. In 1971 the Malaysian Parliament passed the New Economic Policy, aimed at restructuring ownership and enterprise in Malaysian industry so that Bumiputras would gain an increased portion of economic control relative to non-Bumiputras (e.g., Malaysians of Chinese descent) and foreigners. One of the policy's objectives, through the Industrial Coordinative Act, was to shift corporate ownership in the private sector. The previously 10–90 split between Bumiputra and non-Bumiputra businesses was to move toward a 30–70 split instead. Because most trap and small fishing vessel fishermen on the east coast of peninsular Malaysia were Bumiputras, granting compensation for damaged gear or displacement was an obvious way to direct funds toward this economic group. The government, for instance, asked Esso to pay fishermen for damages to fishing traps and their removal during seismic surveying in 1976–78; the company paid U.S. $43,000. The Malaysian government also claimed that pollution from tankers in the Strait of Malacca was destroying fishery resources in the area, which should be prevented or at least compensated. Oil pollution had significantly reduced Malaysian fishermen's catches, affecting about 65 percent of all peninsular fishermen. In this way the government made compensation for fishermen part of its broader national economic program.[120]

Financial Adjustment through Joint Ventures

After 1980 the government was forced to reverse its efforts to limit the level of oil and gas production. A depletion policy, limiting national production, had been instituted in 1980 to prevent the rapid exhaustion of Malaysia's relatively small oil and gas reserves (estimated at 2.84 billion barrels). Production in 1980 was kept to 280,000 b/d, representing the nation's largest commodity export at about 24 percent of that year's exports. However, oil prices began to fall in the early 1980s with the glut on world markets and OPEC's inability to sustain prices by cutting production. The government was left in a bind. Oil revenues were financing 50 percent of the government's development projects, and oil tax revenues were 20 percent of government revenue. The 1982 production level of 297,000 b/d could not offset the country's worsening balance of payments, which showed a $1.1 billion deficit for that year. So in 1983 the government increased oil production by 25.8 percent, compared to a 17.4 percent increase in 1982.[121]

The 1983 increase represented the government's abandonment of its depletion policy to generate 60,000 b/d more than the target of 300,000 b/d needed to offset those losses. Export revenues from crude oil production rose to $3.39 billion and those from LNG, started off at $421 million for 1983. By 1984, at government request, total oil production was 460,000 b/d, which came from production-sharing agreements with Esso Malaysia, Shell Sarawak, and Shell Sabah—respectively 40–45 percent of total national production in peninsular Malaysia, 35–40 percent in Sarawak, and 20 percent in Sabah. Forty-five percent of exports were being shipped to Singapore, and 23 percent went to Japan.[122]

Unlike its Indonesian counterpart, the government in Malaysia reduced its production-sharing and income tax revenues in order to increase exploration by multinational corporations. New oilfields needed to be discovered to sustain higher production levels. The government had to give the multinationals an incentive to intensify exploration. By 1982 Malaysia's tough production-sharing contracts, which had increased to an 80–20 split favoring government, were cut back to only 70–30 with a cost recovery of 30 percent. The government also reduced the corporate income tax to 45 percent. This shift was partly a response to pressures from Shell and Exxon for production-sharing terms that would compare more favorably to the work contracts in Indonesia. The latter enabled foreign companies to keep 100 percent of the oil they produced.[123]

Despite these concessions to the multinationals, the Malaysian government continued to launch new state corporate investments in the 1980s. Rather than insist on exclusive state ownership, the government formed a joint-venture oil exploration company, Petronas Carigali. It was equity-owned 50 percent by Petronas, 42.5 percent by BP, and 7.5 percent by Oceanic Exploration and Development Corp. The company began drilling off Trengganu in 1980. Petronas also formed a joint venture with Société National Elf Aquitaine for oil exploration and production in Malaysia.[124]

By 1983 Petronas had also broken into the multinationals' downstream refining operations while retaining its investments in transportation. The company established two refinery projects, in Malacca and Trengganu. It also created a new state-owned company to replace MISC, the state-owned shipping company, as the owner of Malaysia's five LNG tankers, which had been lying idle in Norwegian fiords until 1983.[125] The new company leases vessels to Malaysia LNG, the joint venture between Petronas, Shell, and Mitsubishi.

In 1984 a change in Malaysian government leadership signaled the potential for a simultaneous shift, toward greater state enterprise investment and yet also toward greater private-sector investment in industrial development. The prime minister appointed a potential supporter of state enterprise as minister of finance to replace the former, more moderate minister. But the aims of the 1985 budget to strengthen foreign borrowing and state enterprise investment had to be reversed by the 1986 budget, which restrained public spending to improve the balance of payments and reduce the government's budget deficit. The prime minister also appointed a supporter of private investment to the Ministry of Trade and Industry. Foreign multinational corporations could thus expect improved conditions for private industrial development through the late 1980s. The obvious way to reconcile simultaneous expansions of state and private investment was to continue the policy of joint-venture oil and gas enterprise.[126]

The Malaysian government pursued a more successful entrepreneurial oil policy between 1968 and 1985 than did the Norwegian, British, and Indonesian states. It extended its ownership control, unlike Norway and Britain, to include international operations. It managed to do so without facing major political and financial opposition from foreign governments and international banks, as Indonesia did. However, the Malaysian government did shift from a wholly state-owned to a joint-venture strategy, incorporating multinationals into its expanding oil and gas enterprises.

The initial period, 1968–76, was characterized by a major government assertion of ownership, revenue, and management control. Petronas was formed in 1974, and soon after the partly state-owned shipping company, MISC, invested in tankers to carry Malaysian liquefied natural gas abroad. Thus state investments in oil and gas production and international transportation gave the state integrated international corporate investments. Production-sharing contracts similar to Indonesia's gave the state 59 percent control over Malaysian oil and gas revenues. Furthermore, the Production Development Act of 1974 gave Petronas a share of management control in all production, refining, and transportation, previously controlled exclusively by the foreign oil companies.

The period between 1975 and 1980, however, was characterized by concessions to the Majors, but of a different sort from those that Indonesia made. The Malaysian government reduced its management control over oil and gas operations by agreeing to set a production ceiling and an export limit that would exclude Malaysia from potential membership in OPEC. At the same time the government agreed to eliminate that section of the Petroleum Development Act which referred to state control of refining and transportation. In exchange the Majors granted Petronas a 65 percent equity share in a new refining venture in Bintulu and a promise of exclusive gas shipment contracts overseas for MISC. Thus the state traded away management control in return for ownership control over refining and international transportation.

Unlike Indonesia, but like Norway and Britain, the Malaysian government also asserted operational control over fishing operations. However, the motive behind this control was not political mobilization of the fishing industry, as it had been in Norway and Britain. Instead, the primary reason for granting fishermen compensation was the government's aim to redistribute economic control within the economy to "native" Malaysians rather than Chinese-Malaysians or foreigners.

The 1980–85 period produced the same effect as in Indonesia, injecting more multinational capital into the Malaysian economy, but through different policy choices. The Malaysian government increased production in order to offset a worsening balance of payments and a growing deficit ($1.1 billion in 1982). But unlike Indonesia, the Malaysian government decreased its revenue control by making concessions on production-sharing terms and corporate income taxes in order to stimulate multinational oil exploration and development. To offset these revenue losses, the government also increased its ownership and associated management control by encouraging jont-venture investments between state enterprises and private oil companies.

COMPARATIVE SUMMARY OF NATIONAL OIL POLICY CHOICES

The Norwegian, British, Indonesian, and Malaysian governments chose a wide variety of policies between 1968 and 1985 to
intervene in national oil industrialization. They claimed a share of
revenue control by collecting taxes, royalties, bonuses, and area fees
on the oil operations of private companies. They asserted *management control* over private operations by requiring production ceilings, production schedules, or the use of national industry and
employment. In some cases governments also imposed *safety and
operational regulations* on oil industry activities, to mitigate the negative effects of oil drilling on other industries such as fishing. Finally, all four governments asserted *ownership control* within their national oil industries by taking a participation share in oil leases or
creating state-owned oil companies. They then shared in production and transport operations through carried-interest, production-sharing, joint-venture, or service-contract agreements with
private oil companies.

In bargaining terms the crucial issue was whether governments
would gain control previously held by foreign multinationals or
concede control to newcomers—private domestic and foreign oil
investors and international banks. Investors new and old, foreign
and domestic, were interested in each of the four types of control
mentioned above. International banks were primarily concerned
that a share of revenue control from taxation be used to guarantee
them an adequate return on their investments or service their
loans.

The Indonesian and Malaysian governments gained during
1968–76 and retained during 1975–80 more ownership, management, and revenue control over national oil and gas operations
than did their Norwegian and British counterparts. Furthermore,
only the Indonesian and Malaysian governments gained those
same kinds of control over international transport operations during 1968–76.

In the 1975–80 period both the Norwegian and the British government gained management and operational control over oil and
gas to pay off private domestic shipping and fishing industries. In
addition, both governments informally had to forego planned in-

vestments in vertically integrated oil company operations to assuage political opposition from domestic competitors. In contrast, the Indonesian and Malaysian governments had no major domestic opposition with which to contend. Instead, they were obliged to find ways to appease multinational corporations unwilling to give up international oil transport operations. The Indonesian government had to cut back on ownership control and shift to revenue control; the Malaysian government conceded less by pursuing joint-venture equity investments with the multinationals.

The final period of financial adjustment, from 1980 to 1985, again showed country differences as governments tried to offset their large public debts with oil income. The Norwegian government relied heavily on external borrowing from foreign banks to continue asserting ownership and management control over oil industry operations offshore. The British government pursued a more domestic financial strategy, trying with some success to sell off its state oil company assets. This effort was intended to reduce the country's foreign borrowing requirements. Having failed to dispose of its assets, the government reduced state holdings in the company to a minority share so that further borrowing would not show up in the public debt figures. In contrast the Indonesian and Malaysian governments resorted to bargains with multinational corporations to offset their financial problems. Indonesia in-

Table 3.2. Comparative state shares of oil industry control 1985, in four countries

Control	Norway	Britain	Indonesia	Malaysia
Ownership[1] (state enterprise)	51%	49%	85–88%	70%–100%[4]
Management[2]	50%	49%	85–88%	50–70%
Revenues[3]	70%	60%	66%	45%

Note. Insufficient data are available to make a percentage assessment of the regulatory effect on operations.

[1]"Ownership." These numbers are based on the percentage of state equity in carried-interest, production-sharing, and joint-venture production and transport operations at the national and the international levels.

[2]Management—the percentage of control by the state according to either state-equity shares in companies or state shares in leases.

[3]Revenues are based on rates of taxation by the 1980s.

[4]Malaysia was the only one of the four countries with 100 percent state-owned control of international tanker transportation, in this case in gas.

creased its share of management control and revenues from the multinationals through production-sharing arrangements, and Malaysia continued to pursue its earlier strategy of joint ventures as a way to favor both state and multinational investment.

But to assess the hypothesis advanced in Chapter 2 that the greater the degree of state autonomy, the greater will be the extent of state enterprise, we need to tally up state gains. Table 3.2 shows the relative size of state shares of ownership, management, and revenues in the four countries by 1985. (The data are inadequate to make a similar assessment of operational control.) The findings show that the Indonesian and Malaysian states have gained significantly greater equity shares of state enterprise in production and transport operations, nationally and internationally, than did the Norwegian or British states. Chapter 4 explains these findings by using the statist bargaining perspective developed in Chapter 2.

CHAPTER FOUR

A Statist Interpretation
of Oil Policy Choices

The Norwegian, British, Indonesian, and Malaysian governments had remarkably similar reasons for creating state oil enterprises between 1968 and 1976. Governments were dissatisfied with the limits on the amount of economic rent they could extract from foreign oil companies, and nationalistic pressures were building for "fairer" shares. Governments encountered limits when they raised corporate income taxes, royalties, or bonuses beyond a minimum profit share acceptable to foreign companies. Companies could declare extraction uncommercial, slow exploration and production, or leave to produce oil elsewhere. Governments used state enterprises to expand their own share of oil wealth and control offshore without subtracting directly from the existing share of the multinational corporations. Foreign companies shifted their profits and control in international trade and marketing beyond the control of the individual state.

But Indonesia and Malaysia differed substantially from Norway and Britain in how their societal groups responded to state oil enterprise. Only when state oil policies in Indonesia and Malaysia began to interfere with international operations did the multinationals use their financial influence to block state investments. And only when domestic companies in Norway and Britain mobilized parliamentary opposition did they curtail state oil company expansion that jeopardized the industry shares of domestic private business.

In this chapter I examine the similarities and differences among the four countries from the statist perspective developed in Chapter 2. I explain the three stages of government entrepreneurship (1968–76), multinational and domestic group responses (1975–80), and financial adjustment (1980–85) discussed in Chapter 3 according to societal constraints on autonomous state interests in oil. This perspective elucidates not only why all four governments formed 100 percent state-owned oil companies but also why private domestic groups constrained state oil enterprise in Norway and Britain more than did multinational groups in Indonesia and Malaysia.

THREE STAGES OF POLICY CHOICE AND ADJUSTMENT

Three stages of government policy choice and actor adjustment characterize oil industrialization between 1968 and 1985. The first stage, the *sovereign entrepreneur* policy choice, had occurred in all four countries by 1976. Governments began to own oil operations. They did so not by directly taking over private operations but rather by acquiring state equity participation in new oil operations. They thus cut into the relative but not the absolute share of the private sector in national oil operations. Governments either increased their equity share in new leases to private companies or created fully government-owned companies to engage in new oil operations. The Norwegian government, for example, first used profit-sharing and carried-interest agreements to gain an equity share in multinational oil operations. However, the government reached a limit on the amount of participation it could gain without rendering oil operations uncommercial from the viewpoint of the multinationals. The government shifted to direct ownership of oil operations by creating Statoil.

The second stage, the *policy payoff* of 1975–80, was more pronounced in Norway and Britain than in Indonesia and Malaysia. Governments in the two industrialized countries made greater concessions of policy-related benefits to domestic groups than did governments in the two less developed countries. Policy concessions to multinational corporations differed in degree but not in general form across the four countries. The period was characterized by government shifts from ownership back to the control of revenues, management, and operations to preserve shares in oil ownership for private-sector actors—the policy payoff for government entrepreneurship in the earlier stage. Major oil companies in Indonesia put up less resistance to tax increases to 85 percent than they did to state corporate monopolies over national and international oil operations. In Norway the government paid off shipping groups by using its discretion over production levels, leasing requirements, and employment stipulations to grant shippers a share of oil-related subcontracting.

The third stage, *financial adjustment,* set in as governments tried to offset unanticipated national debts and balance-of-payments deficits after 1980. Overinvestment in oil and falling demand and prices in the early 1980s had created these financial troubles. Governments were forced to rely more on traditional forms of oil income to gain the large amounts of cash they needed to pay off national debts quickly. They raised corporate income taxes, royalties, and bonuses. The oil glut of the early 1980s pushed production levels and oil prices down as governments scrambled to stabilize national economies dependent upon oil export earnings. Governments, however, were still able to rely on profits from their state oil companies, which survived despite market fluctuations. General contraction of the global oil market simply intensified competition. State oil companies expanded investments and guaranteed markets in government-to-government deals. These efforts aimed to offset national financial crises that affected both industrialized and developing countries during the 1980s.

THE STATIST PERSPECTIVE

The statist perspective outlined in Chapter 2 explains the government's shift to an entrepreneurial oil policy by focusing on the state as a dominant actor. Shifts in market or international political relations are secondary to this focus. The state is the government bureaucracy, including the ministries, state companies, and state banks, as well as representative branches such as Parliament.

Let us take Norway as an example. The statist perspective requires three analytic components. First, the Norwegian state had
an autonomous interest in oil. The government expressed Norway's interest in gaining greater national economic control over
North Sea petroleum by creating a public oil enterprise. The state
acted independently of, and contrary to, private domestic and foreign societal interests that soon began to oppose the policy. Foreign
multinationals never wanted to lose their relative share to the state.
Nor were domestic Norwegian groups willing to give up oil-related
contracts or other industrial opportunities so that the public sector
could make gains in oil. Eventually domestic shipowners and fishermen mobilized through Parliament in the mid-1970s to oppose
government oil policies.

Second, the Norwegian state had the institutional capacity to
sustain autonomous state oil enterprise. Bureaucratic bodies of the
state, such as the Ministry of Oil and Energy and Statoil itself, were
able to perpetuate entrepreneurial policies despite instances of
parliamentary opposition. By the 1980s, in fact, the Norwegian
bureaucracy itself depended upon Statoil as a source of income and
external bank loans to offset the nation's growing public debt and
nonoil deficit. This strength of the bureaucracy relative to representative elements is the second component needed to explain the
survival of Statoil.

Third, the Norwegian state changed in structure as it combined
governance with entrepreneurship. A statist approach analyzes this
as an autonomously generated institutional change. Shifts in the
global oil market or changes in bargaining between the state and
multinational or private domestic companies were important conditions but not sufficient to explain the internal growth and transformation of the Norwegian state. Thus a statist perspective explains the creation and survival of state oil enterprise by combining
an analysis of the state as an autonomous actor with an analysis of
the state's relative ability to bargain with private-sector actors during periods of international market change.

The statist hypothesis suggested that the greater the degree of
state autonomy, the more extensive would be the use of entrepreneurial oil policies. The four cases—actual policy choices in Norway, Britain, Indonesia, and Malaysia—should therefore demon-

strate the following in sovereign or entrepreneurial relations with private-sector companies. The more autonomous the state is in bargaining with societal groups, including domestic and international actors, the greater is the likelihood that the state will make public policy choices that rely primarily on public ownership. Conversely, the more influential societal groups are in bargaining with the state, the greater is the likelihood that the state will make public policy choices that rely primarily on taxation, leasing, and regulation of private operations rather than on direct ownership by public enterprises. The presence of influential societal groups, therefore, can be attributed either to a weakly autonomous state or to societal groups that retain bargaining power despite strong state autonomy.

POLICIES OF THE SOVEREIGN ENTREPRENEUR (1968–76)

Four international trends in 1968–76 suggested that governments could get a better deal from their oil bargains with the multinationals. First, the formation of OPEC in 1960, followed by member agreements to limit production and raise prices, created artificial scarcity conditions in the global market. Prices rose in 1971 and then quadrupled in 1973. The ability of OPEC countries collectively to raise prices by curtailing supplies when global demand was high created a degree of control over the global oil market. The only precedent for such control was the earlier oligopolistic grip of the Majors.

The second trend was the relative loss of control over global markets by the seven Majors—Exxon, Shell, Mobil, Standard Oil of New Jersey, Standard Oil of California, Texaco, and British Petroleum. Their claims on 90 percent of world oil trade, through equity and buyback agreements with governments, began to shrink after 1973.[1]

A third trend complemented this erosion of the Majors' global control. New oil companies proliferated. Many were government-owned; others were independent American, Japanese, and French oil companies expanding operations overseas. This explosion of oil entrepreneurship was partly due to the high profit opportunities and the need for crude oil which resulted from high prices and artificially created scarcity.

A fourth and final trend was the collective search by governments and oil companies for new production opportunities. Discoveries of major oil and gas fields in the North Sea and the South China Sea, and in offshore areas of Mexico, China, and the United States, promised massive new reserves. With the oil pie expanding, the Majors, the Independents, and even state oil companies could all increase their absolute amounts. In relative terms, however, the Majors had to lose for others to gain.

These four trends eroded the position of the Majors in bargaining with governments, culminating in an obsolescing bargain that had begun to emerge in the late 1960s and was in full force by the mid-1970s. Oil-exporting governments had depended upon taxes that multinationals paid on production of crude, the companies' technological capacity, and their access to international distribution networks. But the oil shortage of the 1970s combined with rising global demand to foster government independence. Accumulating oil revenues, competition from Independents to service government needs for technology and expertise, and opportunities to use Independents as substitutes for the distribution networks of the Majors, all enhanced the bargaining power of exporting governments.[2]

But in none of the four countries I have examined did this obsolescing bargain significantly erode the dominant position of the Majors. Esso and Shell remained important Norwegian producers along with such Independents as Phillips, Amoco/Noco, and Petronord (Elf). British Petroleum, Esso, and Shell were the only British producers whose fields had a potential peak production of over ten million tons per year, leaving Occidental, Amoco, Phillips, and Total to take minor roles in field development. By 1974, 44 percent of the offshore production potential in the British sector of the North Sea belonged to the British-based BP and Shell.[3]

In Indonesia, Caltex, the subsidiary of two Majors, Standard Oil of California and Texaco, continued to produce a major share of Indonesian output with its constant 800,000 b/d in 1973–78.[4] In Malaysia two Majors, Esso and Shell, remained the only foreign oil companies in the country until 1978. Thus, although the role of state oil companies and Independents expanded in all four countries, the Majors were difficult to unseat as primary producers.

The best example of the obsolescing bargain in action was in Indonesia. There the state oil company, Pertamina, gained capital and expertise from Japanese Independents, private investors, and the Japanese Foreign Office. This assistance enabled Pertamina to produce Indonesian oil and market it in Japan through an Indonesian-Japanese marketing joint venture, the Far East Oil Trading Company. By 1978, moreover, a Japanese-Canadian Independent, Inpex/Total (with a French parent company), was the largest producer in Indonesia with over one million b/d, which demonstrated the erosion of Caltex's dominance over production in Indonesia.[5] The cooperation between Pertamina and Japanese Independents thus contributed to the obsolescing bargain between the Indonesian government and Caltex.

It is within this context that the states in Norway, Britain, Indonesia, and Malaysia expanded their institutional capacities between 1968 and 1976 to include corporate powers and ownership of oil production and distribution. For states to expand institutionally, they had to safisfy four conditions. First, they had to have sovereign rights and bureaucratic capabilities that could generate institutional growth for oil management purposes. All four governments claimed sovereignty over oil and gas resources inside national frontiers and bilaterally extended that sovereignty to include offshore areas in the North and South China seas. These territorial extensions gave each government ownership and public control over the extraction of oil and gas resources—and thus an opportunity to play an important part in the development of the national oil industry.

Each of the four governments was also strongly unified behind the idea of creating a state oil company. Bureaucracies (including state banks) and parliaments agreed that the formation of a state oil company was a desirable growth of the state as an institution. In Norway, in 1971, the Labor, Conservative, and Liberal parties initially debated the form that state enterprise would take. Labor interests pushed for a strong state role and won. Parliament unanimously approved the creation of Statoil, which was to become a fully integrated, international oil company.[6] A less consensual debate raged in Britain, where Conservatives argued that state enterprise was the path to socialism. Strong Labour party interests, however, eventually generated full support for the creation of the state oil company.

Much of the unity within the Indonesian and Malaysian states derived from nationalism rather than ideological reconciliation. The historical presence of Dutch and British colonialism had created a strong national belief that independent institutions were essential to protect national industries from foreign exploitation. These nationalist sentiments merged with other national concerns about domestic energy, employment, and public control of oil. A strong unity resulted in both states regarding the creation of state-owned oil companies. The formation of Pertamina had the full backing of President Suharto, the Indonesian Army, the Ministry of Mines, the Ministry of Finance, and BAPPENAS. Economists within the last three state bodies were reluctant to support state intervention in oil operations, however, particularly when it jeopardized their own finances and economic control.[7] Likewise, the creation of Petronas in Malaysia had the full support of the prime minister, his Economic Planning Unit, the Ministry of Industry, and the Ministry of Trade and Shipping.[8] In sum, government leaders in all four countries gained support for and legitimized the creation of state oil companies using rationales based on national and public interest in public ownership of oil.

But state enterprise was also a better policy choice than others because it gave states ownership rather than just control over revenues, management, or operations. Ownership allowed government more autonomous bureaucratic control relative to the state's representative branches. Domestic groups could influence the passage of taxation law and apply pressure on parliamentary representatives to attach management requirements to oil policy legislation. Alternatively, multinational oil companies could influence taxation policies by withholding tax payments or threatening to withdraw their operations if taxes took too great a share of private profits. Multinational companies had less influence over management decisions, of course, unless they threatened government bureaucrats with production trade-offs to weaken policy implementation.

Private domestic industrial groups and multinational oil companies had least influence over publicly owned companies. State oil enterprises responded primarily to their corporate directors and secondarily to supervisory bureaucratic agencies. External societal groups could exert indirect influence only by withholding finances or political support through parliament. The most direct form of external group influence required the penetration of state ministries or corporate boards of directors by officials who represented external interests more than the interests of the state in oil industry.

Government leaderships in all four countries not only had sovereign rights, state unity, and bureaucratic autonomy, they were also able to acquire the capital and expertise needed to launch successful state entrepreneurial oil policies. High profitability and high premiums on access to new national oil and gas reserves, mentioned earlier, enabled all four governments to attract the necessary capital and expertise. The Norwegian government had sufficient financial resources from its Treasury and the Bank of Norway until 1978. The British government initially drew from the Treasury's National Oil Account, composed of royalties and area fees from BP and the foreign oil companies working in the North Sea.[9] Because both Statoil and BNOC primarily used participation agreements with foreign oil companies, the expertise of the latter plus that of hired consultants and advisers was sufficient to start oil production.

Nor did capital and expertise come from governments in Indonesia and Malaysia. Instead, Independents and Majors were willing to exchange oil financing and technical assistance for access to new national oil and gas reserves. Pertamina gained full control over production and management, with financial and technical help from Independents in exchange for control over a share of the oil supplies produced. In Malaysia, unlike the Indonesian case, two Majors, Exxon and Shell, were willing to cooperate in production-sharing arrangements for financial, technical, and oil-supply exchanges but only for modified management-sharing. Indonesia and Malaysia were thus drawn into closer financial and technical relationships with the multinationals than were Norway and Britain at the start.

But state oil enterprises needed more than just the initial sovereign claims, state backing, and financial and technical assistance of their governments to survive. Survival in the 1970s and 1980s required that state enterprises possess several institutional characteristics. First, state companies needed charters that prescribed the government interests they should adopt as initial objectives and then use to legitimize subsequent entrepreneurial activities. In all four countries state oil company charters directed state companies to internalize production, transportation, refining, and marketing operations within their own corporate structures by becoming vertically integrated oil companies.

But in all cases state oil companies, as implementers of oil policy and as competitive oil companies, also acquired interests of their own. These interests grew out of initial charters but also branched into new areas of corporate business and politics. Corporate interests beyond the stated charter objectives included the reinvestment of oil profits, borrowing and oil sales overseas, the expansion of market shares, and the creation of subsidiary companies. Pertamina, for instance, veered away from the Indonesian government's original interests, reinvesting its oil profits during the 1970s to expand oil operations rather than channeling those earnings to the Treasury to pay for national development efforts.[10] The British National Oil Company defied the Energy Department's prohibition on foreign borrowing by taking out a loan of $825 million from a group of American banks in 1977. Øystein Noreng has suggested that BNOC leaders saw potential liquidation for the state company if the Conservatives came to power in 1979. The external loan was used to help guarantee BNOC's survival in a Conservative future.[11] Both instances demonstrate that although state oil companies provided oil supplies for their governments, they also engaged in normal business practices. They reinvested earnings and borrowed overseas, sometimes defying government directives when they did so.

State oil companies also enabled states to expand their interests to include greater business shares of vertically integrated oil production, transportation, refining, and marketing. In many cases these interests involved the creation of new state subsidiaries, joint ventures, and other business partnerships. Although government agencies directed state oil companies toward vertical integration, they then began to worry that they would lose control or that such investments might interfere with other foreign policy objectives. Conservative members of the Norwegian Parliament, for instance, became fearful that "the Labor party's baby," Statoil, might attempt to monopolize oil-related shipping, pipeline, and other construction opportunities, excluding other Norwegian private-sector investments from oil-related activities.[12] This same fear of losing control was present in intensified form in Indonesia. Pertamina tried to make ocean-going tanker investments that violated the Ministry of Finance's sense of an acceptable debt-to-asset ratio for new investment. The company's investments also defied the ministry's assessment of the foreign economic policy implications if Pertamina succeeded in carving out a share of tanker operations for transporting Indonesian oil to Japan.[13] In both cases state oil companies derived their investment strategies from initial government directives. But investments, once under way, became perceived by other state agencies as detrimental to broader state economic and foreign investment objectives.

Not only were state corporate policies discordant with other government policies; state oil companies also developed their own managerial interests. These interests grew from opportunities for managing national industrialization through state investment. First, state companies could retain monopoly control over the national share of oil and gas exploitation. Statoil successfully monopolized the national share of Norwegian oil and gas leases by its priority position over two competitor Norwegian companies, Norsk Hydro, which was partly state-owned, and Saga.[14] This monopoly of national shares to the exclusion of nationally owned private oil operations could be observed in Britain, Indonesia, and Malaysia as well.

Once they had a monopoly share of nationally owned operations, state companies could use their political discretion to buy support from private-sector domestic companies. State companies could appease competitors by promising them preferences in subcontracting for drilling, supply, or transportation. In Norway, Statoil bought off opposition from within private Norwegian shipping by informally promising shipowners dominant shares of oil-related shipping and drilling operations. Likewise the Malaysian government appeased its Ministry of Trade by promising that the national shipping company, MISC, would obtain a monopoly over the transportation of Malaysian gas if it invested in liquefied natural gas tankers.

State oil companies developed other managerial interests that stemmed from their sectoral wealth and power relative to shipping, manufacturing, and other domestic industries. These industries were in decline either because of lack of demand within their sectors or because of oil-induced inflation or currency appreciation. State oil companies were able to use oil revenues to guide national planning by subsidizing or stimulating investment in industrial sectors of their choice. In Norway and Britain, Statoil and BNOC used oil revenues to subsidize declining or nationalized public industries such as shipbuilding.[15] In Indonesia, Pertamina used oil revenues to expand shipping and construction sectors through Pertamina subsidiaries and even to build hospitals, schools, and roads. However, these latter investments in public health, education, and infrastructure were met with great opposition from the national government's planning agency, BAPPENAS, which was itself responsible for such investments. It was at this point in Pertamina's multisectoral expansion that the foreign press began referring to the company as a state within a state.[16]

Thus state oil companies in all four countries developed new corporate and managerial interests. These interests came from, but also were autonomous from, the interests that governing coalitions had specified in the companies' initial charters.

The final characteristic necessary for state oil companies to survive was institutional autonomy. State companies needed an independence within the state itself sufficient that ministries of finance, departments of energy, and other bureaucratic agencies could not subordinate them. Although both the Norwegian and the British state oil companies were initially placed under energy ministry supervision, they became increasingly autonomous in their corporate decision making. Investment decisions were based more on market pressures and reinvestment strategy than on changing ministerial policy objectives. The Norwegian government placed Statoil under the Petroleum Directorate's supervision specifically to avoid such delays at the political level of the ministries. In Britain BNOC turned the tables by exerting its own influence over the Treasury's policies once state oil company revenues became an important share of the Treasury's purse.[17]

State oil companies in both Indonesia and Malaysia had even greater institutional autonomy. Pertamina was given production, operational, and financial control over its own decisions from the start. The state oil company was thus able to reinvest its earnings without channeling them through the Indonesian Treasury and to take out foreign loans, despite ministerial efforts during the early 1970s to stop both practices. But neither the effort of the Ministry of Finance to get foreign oil companies to pay the Treasury directly nor Parliament's effort to prohibit the state oil company's long-term loans from foreign banks eroded Pertamina's autonomy in procuring finances.[18]

In Malaysia the autonomy of the state oil company was more clear-cut. The prime minister's Economic Planning Unit worked closely with company leadership to further Petronas's investments in the production and transportation of oil and gas. Although not autonomy from prime ministerial control, this tight link enabled oil and gas policy to develop without constraints from other Malaysian ministries.

In sum the two developed countries, Norway and Britain, and the two less developed countries, Indonesia and Malaysia, made similar policy choices to expand public ownership of oil. All four governments formed new 100 percent state-owned oil companies even though they had previously held shares in private oil companies. Through state companies governments enhanced their control by creating corporate functions that had implicit managerial and bargaining powers associated with state investment strategies. From a statist perspective, the emergence of a bargaining limit, despite conditions suggesting that the bargain between states and multinational corporations was obsolescing, created environmental conditions needed for the expansion of the state. Other institutional conditions—sovereignty, state unity, bureaucratic leadership, capital, expertise, and government authorization— were necessary for state oil companies to be created and given bureaucratic autonomy. Eventually state oil companies came into conflict with the interests of other state agencies over financial or policy objectives. Once states expanded to include autonomous state oil companies, it appears, they were likely to face internal bureaucratic conflict, particularly if these new companies depended minimally upon inputs from other bureaucratic agencies or external societal groups.

Differences between developed and less developed countries show up in the extent to which state oil companies were vertically integrated. The Norwegian and British states invested primarily in production, domestic refining, and pipeline transportation. In contrast the Indonesian and Malaysian states invested not only in production and domestic refining but also in domestic or international transportation and marketing. We cannot explain this greater vertical integration by less developed countries merely by referring to the institutional growth characteristics of the state. Explanation of the difference also involves the bargaining relationships between state, multinationals, and domestic companies. That bargaining would impose limits on state entrepreneurship between 1975 and 1980.

POLICY PAYOFFS (1975–80)

Between 1975 and 1980 differences between the developed and less developed countries began to appear in the international and domestic bargains that states achieved. State oil companies in Indonesia and Malaysia made two-way bargains with multinationals, but those in Norway and Britain created three-way bargains, granting domestic groups a share. In the two-way bargains in less developed countries, public ownership was traded for increased government oil revenues. The three-way bargains in developed countries required trade-offs in terms of public ownership as well as increases in management and operational control by government. The latter increases were needed to carve out a share of oil benefits for domestic shipping, fishing, and labor, groups that had not figured in the state-multinational bargain made earlier.

We need to subdivide the concept of the state if we are to explain from a statist perspective these reunified governments and reworked bargains. Essentially, the period 1975–80 demonstrates that the bureaucracies of the four governments were dominant over parliaments in oil matters. Not only were bureaucratic agencies able to reunify the state by forcing state oil companies to accede to their authority, but they were also able to prevent the dissolution of state oil enterprises. Bargains were reworked despite strong challenges from parliaments and even, in the case of Britain, the prime minister.

Reunifying the State

The managerial role of bureaucracies was strengthened between 1975 and 1980. The increased authority to collect and disburse oil tax revenues, manage subcontracting and employment levels through leasing policies, specify safety regulations for oil operational procedures, and carve out shares for state oil companies contributed to pervasive administrative control. Bureaucracies could act as corporate competitors, industrial leaders, clearinghouses for contracts, channels for redistributing oil revenues, and caretakers of national employment.

These managerial roles were played primarily by the Ministry of Oil and Energy, the Petroleum Directorate, and Statoil in Norway, and by the Department of Energy and BNOC in Britain. Managerial control in Indonesia was primarily in the hands of Pertamina and the Ministry of Mines; in Malaysia it was concentrated in the strong bond between the prime minister's Economic Planning Unit and Petronas. Each of these managerial partnerships combined bureaucratic authority with corporate power in pursuing oil policies or bargaining with multinational corporations or domestic groups. Essentially this bureaucratic-corporate strength enabled bureaucracies to use administrative decision making to make policy choices. This strength grew as states in all four countries became increasingly intertwined in their oil industries.

Bureaucracies had the institutional capabilities to intervene in private oil operations (consistent with Weber's analysis of the rational-legal state). They were able to use their compulsory jurisdiction over oil matters, their own organizational framework of administration, and their sovereign rights to enforce policy decisions. With such capacities, states had the potential to play significant roles as managers at an advanced stage of oil industrialization. State oil enterprise and bureaucratic administration offered stability, constancy, predictability, and the possibility of national planning.

In contrast to these strong oil bureaucracies, representative state bodies such as parliaments were in relative terms weakened by their primary roles of legislation and approval. They could alter national levels of oil production, legislate uses of national employment or national contractors, or approve state oil company loans. But these abilities were generally not effective in curtailing the administrative and entrepreneurial choices of oil bureaucracies. Furthermore, parliamentary actions were constrained by societal interests. They thus vacillated with electoral swings, changes in the salience of political issues, and partisan shifts.

Once the managerial apparatus of the state had become integral to national oil industries, state leaders were hard pressed to downplay their role. The British and Indonesian cases are particularly dramatic examples of how governments were wedded to their state oil enterprises, willingly or not. Consolidation of the state was also essential because powerful state enterprises could be neither eliminated nor tolerated as components of the state potentially more powerful than the bureaucracies of which they were a part.

Prime Minister Margaret Thatcher tried to sell off BNOC's profitable oil production and trading operations entirely; her effort to eliminate state oil enterprise within the state was both a partial success and a partial failure. It was a success in that she reduced state ownership to a minority share, enabling the company to bail out the Treasury with foreign borrowing without increasing the nation's reported public-sector borrowing. But it was a failure in that Thatcher could not entirely divest the company as she had intended. The Treasury and other domestic industries, as well as the Conservative party, were financially or politically dependent on it. The Treasury needed the long-run profits and ability to earn foreign exchange. Nine national industries depended upon oil subsidies; oil rig, platform, and supply boat construction; subcontracting; or employment generated by BNOC or by the Department of Energy's stipulation that British goods and services, when competitive, be used in the North Sea. These roles in management and subsidization played by the alliance of BNOC and Department of Energy were hard to replace by a simple sale of BNOC and a Treasury bail-out of declining British industries with the cash. Furthermore, the indebtedness of the state oil company to a consortium of American banks made liquidation of the company even more difficult for the Conservatives. Despite six years of Conservative efforts (through 1986) to privatize the company completely, Britoil has survived, testimony to the resilience of state oil enterprises as permanent elements of national oil policies even where governments hold only minority shareholdings. Prime Minister Thatcher had no choice but to embrace the state oil company after she found that her attempt to dissolve the company could only partly succeed.

The Indonesian president's need to subordinate Pertamina to the control of the Indonesian state was more a defensive measure than an unwilling accommodation (as in Britain). Pertamina had extensive managerial power within the Indonesian oil industry as well as over army finances and over the regional development of schools, hotels, hospitals, and roads. This power ended, however, once foreign banks called for debt payments and the Indonesian Army retracted its support because it was no longer receiving side-payments. The Indonesian president was forced to bring Pertamina back under the control of the government bureaucracy. Had the army continued to support Pertamina, any effort to control the company—by the president, Ministry of Finance, or International Monetary Fund—even with the strong backing of the U.S. government, would have been difficult.

President Suharto reconsolidated the Indonesian state by dismantling Pertamina's leadership and organizational structure and divesting part of the company's operations. Weaker oil management could preserve Indonesia's national and public interests in oil development but do so under the auspices of the existing state leadership. This reconsolidation of the state preserved the international legitimacy of Suharto's regime in the eyes of the IMF and foreign governments. It also recreated state unity in the Indonesian bureaucracy, including the presidency and ministries of Finance, Mines, and Planning, the army, the Parliament, and Pertamina.

These two cases demonstrate the strength of state bureaucracies and the resilience of their state oil companies. But that strength was also kept in check by the ability of multinational and domestic groups to use more representative state organs and even bureaucratic agencies to influence state policy.

Better Bargains for Multinationals and Domestic Groups

During 1975–80 international political and economic conditions set the stage for checks on state oil policies. International prices on the spot market had reached record highs by 1975, following the OPEC price rises of 1971 and 1973. By 1978–79 prices had again doubled. This presented not only the seasoned Majors but also the Independents and state oil entrants with tremendous profit potentials. But the Majors were losing ground: state and Independent oil company production in many countries had begun to erode their previous dominant control over the production and marketing of oil supplies. By 1980 the Majors controlled only about 55 percent of global oil trade, while newcomers claimed 45 percent through direct or spot market oil sales in processing deals.[19]

Global shipping was in steady decline. High profits up to 1972 had created high expectations among Norwegian and British shipowners about industry growth. But the glut that appeared in world shipping thereafter caused problems for Norwegian and British shipping companies. Many Norwegian owners attempted to diversify into oil production and transportation; British shipowners were deterred from similar attempts by the dominance of multinational oil companies already owning a share of British shipping and oil operations. Because Indonesia's and Malaysia's nontanker fleets were smaller and restricted to domestic coastal or interisland trade, they were less affected by the decline in the international shipping market.

Fishing in the North Sea was also in a state if not of decline, then of chaos. The decline of the herring fishery during the 1960s, and subsequent government efforts to use quotas to preserve stocks, had created a flurry of fishing activity. Danish, Norwegian, British, French, and other North Sea fishermen competed fiercely for a share of the quotas. Furthermore, British fishermen prohibited from fishing Iceland's waters had returned to the North Sea with hopes of recapturing their previous catches. So although total catch levels were rising, levels for specific stocks were regulated and competition among fishermen was intense. These fishermen were already cramped for trawling space between oil rigs and were suffering from catches reduced when nets were damaged by leftover oil wellheads and other debris. Further interference from oil activities was likely to anger them.[20]

These shifts in oil, shipping, and fishing catalyzed international and private domestic pressures for governments to concede better bargains. The International Monetary Fund and the U.S. government wanted oil-producing governments to conform to the rules of an international monetary and oil regime that would assure the continued dominance of the Majors. The IMF also wanted governments to conform to accounting procedures consistent with the rules of international finance. And the U.S. government was trying to preserve flows of oil to the United States, Japan, and Western Europe. Domestic private sectors, particularly shipowners and fishermen in Britain and Norway, wanted governments to offset their losses in shipping and fishing markets with shares of oil operations or revenues from the North Sea.

Multinational and domestic groups with pivotal power could influence the oil policies of strongly managerial oil bureaucracies. International groups were able to alter policies either by jeopardizing the finances of bureaucracies (Indonesia) or by delaying expensive contracts (Malaysia). American banks dealt a critical blow to Indonesia's entrepreneurial oil policies when they withheld additional loans and refused to renegotiate existing loans to the already heavily indebted state oil company. Similarly, the Majors softened the Malaysian government's bargaining position on production sharing by delaying contract negotiations. The delays caused Malaysia to incur heavy financial losses from its unemployed LNG tankers.

These pivotal exercises of bargaining power by international groups led to revenue shares replacing state equity, in the form of production-sharing or joint-venture agreements in both countries. The outcome was more consistent with the demands of international actors and constituted an adjustment of the two-way bargain between multinationals and states. Essentially, pivotal power had undermined the managerial power assumed by strong government oil bureaucracies.

Multinational groups in Norway and Britain, by contrast, were able to rely on private domestic groups to curtain strongly entrepreneurial state oil policies. Private domestic groups—shipping, labor unions, and fishing—applied pivotal power by withholding parliamentary votes critical to state oil company finances or state oil leasing policies. Representative branches of the state thus acted as channels through which private domestic groups could gain oil-related benefits from state oil policies. Such pivotal checks forced both the Norwegian and the British governments to increase their management and operational control over North Sea oil to make concessions to shipowners, unions, and fishermen. Shipowners gained a share of oil-related operations; unions gained shares of North Sea employment. Fishermen gained delays in oil leasing (in Norway), protective safety provisions, and compensation for equipment damage. Essentially, this pivotal influence of domestic groups transformed earlier, two-way multinational-state bargains into three-way bargains. Domestic groups were now included, either with a share of oil operations or with safety provisions and financial compensation for oil-related losses.

In sum, the two developed and two less developed countries were similar in that state oil enterprises retained major shares of control within national oil industries. This result is attributable to the strong managerial role that government oil bureaucracies and state oil companies played over the 1975–80 period, despite some efforts in Britain and Indonesia to get rid of or tame the state company.

The main difference lay in whether it was bureaucracies or parliaments that curtailed strong oil bureaucracies. The pivotal financial power of international groups forced bureaucracies in Malaysia and Indonesia to rely more on oil revenues than on ownership in their two-way bargains with multinational corporations. In contrast, the pivotal parliamentary power of private domestic groups forced governments in developed countries to rely more on managerial and operational control of oil. Such administrative discretion carved out domestic shares in three-way bargains between state, multinationals, and domestic groups.

The result was that the two less developed states, Indonesia and

Malaysia, could be more autonomous as entrepreneurial leaders of national oil economies than could the two developed states, Norway and Britain. Strong domestic groups in Norway and Britain not only penetrated the parliaments but also wielded partisan influence through a politically mobilized electorate. The colonial origins of Indonesia and Malaysia made them less vulnerable to domestic groups, which except for Chinese-Malaysians remained relatively underdeveloped in both political and financial terms.

But though the two developed states were weaker than the two less developed states in bargaining with domestic groups, LDC states were weak in bargaining with international groups. The centralized states of Indonesia and Malaysia acted as funnels for foreign investment capital from multinational corporations and international banks, and aid from foreign governments. This role made them less resistant to foreign influence, particularly in the case of Indonesia where the Ministry of Finance badly needed foreign investment capital, in part because of Pertamina's control over oil revenues.

In the Indonesian case the government shifted from depending upon multinational corporations for income to internalizing coalitions of advisers who represented foreign capital groups (including multinationals) within state agencies. The coalition within the Ministry of Finance included foreign-trained and foreign economists, giving legitimacy to the influence of international groups. In this way the interests of external groups penetrated the state and offset its ability to pursue autonomous oil interests.

Strong states in less developed countries may thus be penetrated by outside international financial interests (linked with domestic groups) in oil exploitation. But even this internalization of private international interests did not overwhelm the ability of Malaysia and Indonesia to bargain for their own oil interests in early stages of national oil operations. The two developed states had a harder time bargaining with strong domestic groups also represented within state organs.

But the bargains of the late 1970s could still assume that the oil pie was expanding. Circumstances after 1980 demonstrated that expansion is not always a good thing. If the pie is too large, vendors may not be able to sell their pieces.

FINANCIAL ADJUSTMENT (1980–85)

The 1980s began with a major contraction in the global economy, a contraction due in part to the oil glut and oil-induced inflation created by overproduction in the oil crisis years between 1971 and 1979. The contraction in the global oil market took two forms: a permanent restructuring of the oil market to the favor of state companies, and a decrease in demand for crude and refined oil products.

By 1985 a quarter-century's restructuring of the global oil market had clearly favored state oil companies and Independents over the Majors. The dominance of the Seven Sisters over 90 percent of global oil trade, a commonplace before 1973, was at an end. By 1982 the Majors owned only one-tenth of the noncommunist world's oil reserves and produced less than one-fifth of its oil. It was only by buying supplies from nationalized oil companies that the Majors were able to trade two-thirds of the noncommunist world's petroleum products in that year. Compared to 1980 levels, furthermore, the Majors' global refining capacity was down 30 percent, oil tanker fleets were down between 30 percent and 50 percent, and petrochemicals were down 40 percent.[21] This general shrinkage of the Majors' share of the oil market forced those companies to innovate.

By 1983 the Majors were launching two new strategies to save their stake in the global economy. The first move was to diversify into nonoil investments. Exxon bought Reliance Electric, Gulf established a nuclear alliance with Shell, British Petroleum began to invest in metals, and Mobil bought Montgomery Ward.[22]

The second move was to decentralize their profits. The Majors separated their refining and production operations in response to the high prices they were having to pay OPEC countries for over half their supply of crude oil. Refining was separated from production so that the costs of each operation could be balanced separately; this was especially needed because production in many countries was being taken over by national companies. Decentralization of profit also addressed the Majors' need to give their foreign subsidiaries more bargaining flexibility. It enabled subsidiaries to negotiate taxes and regulations with host governments without being subject to the tight control of central corporate head-quarters. Both diversification and decentralization thus aimed to protect profitable areas of operation, by expanding into nonoil investments, by distancing costly refining from production, and by increasing the bargaining flexibility of subsidiary operations in different countries.[23]

State oil companies quickly filled the market positions left vacant by the Majors. The fifteen largest Western oil companies were down in share by 60 percent between 1973 and 1983, and state oil companies had been able to increase their share to 42 percent of global oil trade. Furthermore, half of OPEC sales had shifted to these state oil companies and to other, Independent oil companies that were not among the world's fifteen largest. By 1983, in sum, at least half of global oil trade was between OPEC governments, state-owned oil companies, and medium-sized Independents.[24]

What is particularly important is that by 1980 the state-owned oil companies in Norway and Britain were very profitable. Statoil reported pretax revenues of $49 million; BNOC reported a new profit of $339 million.[25] This profitability made it more difficult for Conservative governments in both Norway and Britain to justify the sale of nationalized oil interests.

Despite structural shifts in the composition of the global oil market, that market was shrinking. Between 1980 and 1982 noncommunist world oil demand fell by an average of 5 percent per year, and the International Energy Agency projected falling demand into 1985. Even more startling, the demand for oil in the year 2000 was projected as slightly less than demand in 1979 (31 million b/d in trade), because of the effects of conservation measures and shifts to coal and natural gas by most national governments.[26] Thus the world oil market by 1985 had shifted from the scarcity of the 1970s to a glut compounded by falling demand.

OPEC countries had no choice but to cut production levels. The dispute over production levels in OPEC, however, was over amounts of cuts as well as shares of total OPEC supplies. Countries such as Nigeria were forced by large foreign debts and spending commitments to cut their oil prices. The Saudis aimed to retain dominance in OPEC, high price levels, and price unity among members, but those aims could be dashed by single members of the cartel. Indeed, this Saudi dependence on OPEC cooperation gave weaker members a powerful bargaining chip regarding the size of the Saudi share of OPEC production. Global constraints and intra-OPEC bargaining resulted in a cut in OPEC production levels, to 13.5 million b/d in March 1983. This level retained the price unity that the Saudis wanted, but at a price. The Saudi share of total production was reduced to only 3–5.5 million b/d.[27]

State Adjustment Strategies

The governments of Norway, Britain, Indonesia, and Malaysia were forced to increase their control over oil revenues in order to meet the financial exigencies that a contracting global oil economy imposed on their domestic economies. Government treasuries in all four countries were faced with domestic economic problems carried over from the 1970s: oil-related expansion, oil-induced inflation, and decline in industrial sectors such as steel and shipbuilding. Most of the Norwegian government's oil income was committed to heavy subsidies for shipbuilding, fishing, farming, and textile sectors that were in decline because of oil-induced inflation and currency appreciation. Total subsidies reached $14 billion by 1980. Similarly, British oil gains were eroded by nonoil budget deficits and the need to subsidize unemployed workers and declining industries. Heavily indebted to foreign banks, the government was also compelled to reduce its public-sector borrowing requirements.[28]

The Indonesian and Malaysian governments were similarly concerned about reducing deficits and sustaining the oil income that promised to finance economic development plans. The Indonesian government retained a 70–75 percent share of its foreign exchange from oil and gas exports for economic development. But in 1982 it was forced to reduce its subsidization of domestic fuel prices to offset losses on corporate income tax.[29] The Malaysian government was also financially troubled. Shouldering a $1 billion deficit by 1982, it could not afford a cut in taxes or profits from oil and gas sales. Oil revenues were financing 50 percent of government development projects, with oil taxes providing 20 percent of the total.[30]

Despite the similarities in their financial problems, governments adopted different but not mutually exclusive adjustment strategies to keep national economies afloat. The strategies were of three basic types: 1) borrow from foreign banks, as in Norway; 2) sell off public enterprises, as in Britain; and 3) rely on the multinationals to shoulder some of the financial burdens, as in Indonesia and Malaysia. The last strategy sustained the traditional bargaining between state and multinational corporations in the two less developed countries. But a greater reliance on international banks and the private sector indicated new strategies of public finance for Norway and Britain.

The Norwegian government adopted the first of these three strategies and expanded its oil and gas operations in the 1980s by borrowing heavily from foreign banks. It had incurred a public debt of over $7.2 billion to cover costs and delays in North Sea oil development. Some effort was made to use the 1982 surplus in the balance of payments to reduce overseas borrowing and replace it with financing from domestic credit.[31] The government also continued to rely on its state oil company, Statoil, to push oil expansion forward into the mid-1980s.

In dramatic contrast the British government took the second tack and sold off shares in its public enterprises in a massive privatization. The purpose was to secure immediate cash for the Treasury to finance current spending. Total privatization of oil and gas operations produced $2.4 billion, mostly from sales of BP shares, by December 1984. But by 1985 the Thatcher government had managed to sell off only $741 million of Britoil (formerly BNOC) to the private sector.[32]

Unable to dissolve the state oil company entirely, the British government was forced to modify the aims of its privatization policy. It had to be satisfied with reducing the state's shareholding in Britoil from a majority to a minority position. But this was itself very useful. Minority state shareholding guaranteed that Britoil's foreign loans would *not* henceforth show up in the Public Sector Borrowing Requirement, which the government was trying to reduce.

Both the Indonesian and the Malaysian governments took the third strategy, relying on multinational corporations to meet their financial needs rather than borrowing from banks or selling off state assets. The Indonesian government renewed Caltex's production contract, insuring that the company would remain a dominant producer in the country. But it also compelled the company to absorb two-thirds of Indonesia's OPEC-required production cuts and to invest extensively, to the tune of $3 billion, in new exploration and production. The government made up for its own financial losses from the production cut by adjusting terms in its production-sharing contracts. The government's take rose from the 85 percent typical of earlier production-sharing contracts to 88 percent.[33]

The Malaysian government similarly relied on the financial contributions of the multinationals to continue its expansion of oil activities. But instead of demanding more favorable production-sharing terms, the Malaysian government set up joint-venture investments with foreign oil companies. Joint ventures offered a means for the government to continue to stimulate private investment yet also to continue foreign borrowing and state investment.

Debt Crises and Financial Bargains

Two questions must be answered to understand the 1980–85 period. First, why did the Norwegian, British, Indonesian, and Malaysian governments all continue to rely on state oil enterprises when they were suffering from national and global economic contraction and needed fast cash? Second, why did governments pursue three different adjustment strategies in trying to meet their financial needs?

A statist explanation delves into financial dependencies within the state and bargains reached with external financiers. Governments continued to rely on state oil enterprises because they had in financial terms few alternatives. The 1968–76 period enabled autonomous oil bureaucracies and state oil companies to expand because global oil demand was increasing, production opportunities were numerous, and international banks were willing to make risky capital investments in oil and gas. The 1975–80 period was characterized by external political and financial pressures on thriving government oil bureaucracies. Domestic industrial groups, international banks, and multinational companies worried about losing their relative share of petroleum wealth. By the 1980–85 period, global economic recession had turned oil-indebted producer governments into precarious financial entities. States suffered tremendous internal fiscal strains. Government treasuries in Norway, Britain, Malaysia, and Indonesia shouldered enormous public debts and ran economic deficits, which, in cases such as Britain, substantial oil and gas revenues temporarily hid. Under fiscal pressure, treasury departments looked for sources of financing. State oil companies were still profitable in all four countries, and government treasuries were therefore compelled to depend upon them as sources of finance. State oil companies provided the state with income to supplement taxes and royalties and as channels through which to increase public borrowing. The case of Britain is particularly interesting. The sale of Britoil shares in the early 1980s was supposedly aimed at privatizing public enterprise. Its effect was actually to increase the government's capacity to incur public debt that would not appear in the government's accounts. Raymond Vernon also makes the point that ministers could use the state oil company as a strategic cache of resources that were "unavailable or unknown" to public and Parliament.[34]

This contraction of the state, caused by the internal financial dependence of treasuries on wealthy state oil enterprises, was particularly apparent in Britain and Indonesia. In Britain the Treasury insisted on retaining control over Britoil's foreign borrowing limits. It also demanded that all of the company's surplus cash be transferred directly to the Treasury rather than loaned out at interest. The Treasury went one step further. It dismissed corporate leaders who disagreed with government policy and exercised full authority over sales of corporate shares and over accounting procedures.[35]

Much of this financial control had already been established in Indonesia between the Ministry of Finance and Pertamina during the 1974–75 reorganization of the company. Further clamps, however, were placed on the company in the 1980s. By shifting a Pertamina director to head the Ministry of Mines and Energy, President Suharto recreated the strong link between the ministry and the state oil company. (The same link had existed in 1966, when

Ibnu Sutowo was both minister of mines and head of Pertamina's predecessor.) In the 1980s, however, the Ministry of Mines and Energy specifically instructed Pertamina to maximize foreign exchange earnings through exports and to channel its oil revenues to the Treasury in order to finance national development plans.[36]

Thus the reconsolidation of the state during the 1980–85 period directly resulted from the fiscal dependence of poorer state bureaucracies on wealthier state oil enterprises. Bureaucracies may have been relatively poor, but they had the authority to subordinate these enterprises to their wishes. Fiscal strains led to a reduction in the autonomy of state oil enterprises, assuring government treasuries of the oil finances they needed.

The second question is why governments adopted different strategies to solve their financial problems. We can answer it by examining bargaining patterns in these reconsolidated but financially troubled states. Essentially, the choice to borrow from banks, sell off state assets, or rely on multinationals reflected a new three-way bargain for states. The previous three-way bargain, between 1975 and 1980, had included domestic industrial sectors in the benefit package. The new bargain that emerged in the early 1980s included international banks as a third partner with states and multinationals. By the beginning of the 1980s international banks had already recycled $324 billion, the total foreign assets of OPEC governments, in global investments. They had a large stake, therefore, in seeing that governments continued to service their debts despite the financial difficulties characteristic of the early 1980s.

Debt crises occurred for oil producers when the rapid oil-induced industrializations of the 1970s collided with the declining demand of the 1980s. Peter Cowhey has pointed out that the long-term crisis for oil exporters did not actually set in until 1982. At that time the previous pattern, oil price rises followed by falling oil demand and then slow stagnation and upward movement of prices, could no longer eliminate the vast petrodollar surpluses created by higher prices.[37] States were left dependent upon international commercial banks for debt financing to expand state oil enterprises and other public investments. Cooperation with multinational oil companies continued to be important for all four countries to retain levels of production. But the state's share of financing came increasingly from foreign indirect investment by international banks rather than from foreign direct investment by the multinational corporations themselves.

Such debt-financed industrialization did not only occur in less developed countries, which accounted for 57 percent of total external public debt committed by international financial markets.[38] The Norwegian and British governments also contracted extensive loans from foreign banks to finance state oil expansion as well as general economic development. For instance, Norway had a 3:1 ratio of nonoil to oil external borrowing in its total $27 billion public debt in 1980.[39] The same ratio holds for Indonesia's slightly smaller overall public debt, $24.5 billion by 1980, of which $6 billion was attributable to the state oil company's debts.[40] Both developed and less developed countries were thus heavily reliant on debt financing by private financial institutions for nonoil and oil investment.

But the indebted industrializations of Norway, Britain, Indonesia, and Malaysia were not acknowledged internationally as were the financial crises that occurred in, for example, Brazil. By 1980 Brazil's $30 billion external public debt was only slightly greater than that of Norway.[41]

This happened even though the international banks had taken a substantial role in financing the state expansions of the 1970s. Those banks thus had a major stake in the economic contractions that occurred in the early 1980s. Government income from oil exports, reduced by production and price cuts, would not cover foreign currency requirements generated by imports and thus threatened to worsen balance-of-payments deficits. Governments thus became less able to service their economic development loans from international banks. This reduced ability to service debts explains why government treasuries increased their administrative and legal claims on the revenues of state oil companies. Treasuries needed to guarantee that the dominant source of autonomous state income—oil—would be used for general economic purposes.

Financial adjustment strategies thus took different forms—borrowing from banks, injecting private finance into state operations, relying on traditional multinational investment. But in the early

1980s the name of the game was the same: capital. Both international banks and multinational companies had it; unlike in the 1960s, however, state oil companies had it too. International banks demanded debt servicing, the multinationals demanded oil production shares, and indebted government bureaucracies demanded more foreign exchange from their state oil company's oil exports. A new three-way bargain was struck.

THE STATIST HYPOTHESIS

The evidence through 1985 confirms the statist hypothesis. The policy outcomes, as we have seen, confirm that autonomous, centralized bureaucratic regimes—Indonesia and Malaysia—made greater entrepreneurial gains in both national and international oil and gas operations. These governments encountered less domestic opposition than the Norwegian and British states from shipping and fishing groups. The Indonesian and Malaysian governments also made the greater gains in sovereign control, acquiring significant shares of revenue and managerial control within the industry. These two less developed countries made gains in oil industry control in the face of consistent opposition from multinational corporations and international banks intent upon retaining their share of international oil operations.

More representative regimes, in Norway and Britain, enabled domestic groups to gain larger shares of both sovereign largesse and entrepreneurial concessions. Though the multinationals retained a substantial share of oil operations, international banks actually gained some control over oil revenues. But bank inroads were made only after the two governments had incurred substantial oil debts through external borrowing. These concessions to domestic and international groups reduced the overall revenue, management, and ownership control left for governments. However, the Norwegian and British states did make greater gains than the two less developed countries in operational control regarding industries, mainly fishing, affected negatively. In sum, Indonesia and Malaysia achieved more as Sovereign Entrepreneurs than did Norway and Britain.

In this chapter I have argued that a statist perspective can explain the policy choices made by the Norwegian, British, Indonesian, and Malaysian states during the oil industrialization period, between 1968 and 1985. This perspective focuses on the bargaining consequences of shifts in state structure. Oil opportunities in 1968–76 enabled states to become more autonomous from both domestic and international societal groups. This separation of state and society was the basis for a new bargain between states and multinational corporations, a bargain in which the ownership control of oil was repossessed by states acting as public corporations rather than merely as sovereigns.

The years 1975–80 were a concessionary period for paying off opponents to state oil expansionism. States that had both governing and entrepreneurial capacities posed significant threats to private-sector competitors for oil market shares. These capacities also enabled states to create new bargains that cleared the way for state ownership. States conceded some management and operational control to appease multinationals and the IMF. In this period the distinctive political features of developed and less developed countries showed up in policy choices. Representative bodies of the state in developed countries enabled private domestic groups to gain more management and operational control of oil activities. More centralized bureaucracies in less developed countries provided the multinationals with access to decision making about ownership and revenue control of petroleum.

In 1980–84 the state suffered fiscal strains caused largely by contractions in the global oil market. These internal fiscal strains pitted indebted treasuries against wealthier state oil enterprises. The political and administrative claims of the former reconsolidated the state as a unitary actor despite financial asymmetries within the state. This overriding authority of state treasuries guaranteed the bargain with international bankers: increased revenues from oil profits and taxes would service the debts owed to foreign banks.

CHAPTER FIVE

The Sovereign Entrepreneur Model: Implications for Other Countries

What does the growth of state-owned oil companies imply for other countries? Have state-owned oil companies elsewhere also acted as corporate arms of the state, claiming national shares of oil industry profits and industrial capacity? Have multinational oil companies reacted by decentralizing their profit shares in international trade and shipping networks among countries while still competing for national shares? Finally, have domestic interest groups demanded that the state concede either oil industry–related work or financial compensation when they were displaced by oil industry activities? If these three trends regarding oil industry organization can be observed in most capitalist countries, then we may expect patterns to appear in the interaction of state, multinationals, and domestic groups which resemble those in Norway, Britain, Indonesia, and Malaysia.

I shall argue in this chapter that most capitalist states are now involved in the oil business; this is the basis for a *sovereign entrepreneur model* of state behavior in national oil industries. The model emphasizes the advantages that can be but are not always achieved by combining sovereign powers and entrepreneurial capacities. Sovereign powers, discussed in previous chapters, include the ability to make unilateral territorial claims, to legitimize state ownership, and to demand that oil lease agreements be renegotiated to include a share for state companies. The state's entrepreneurial capacities include the ability to make investment decisions and to compete in business as a government entity. State enterprise can also use low-interest government loans to buy technologies and expertise for exploration, production, and transportation.

The sovereign entrepreneur model involves two major patterns, which appear individually or together in the four countries analyzed in this book. The *domestic concession pattern* results from the presence of domestic opposition groups strong enough to demand concessions from the state. The *multinational concession pattern* results from the presence of multinational opposition groups (including corporations, international banks, and foreign governments) strong enough to demand state concessions.

In both concession patterns the degree of public ownership of the state oil company and the extent of its vertical integration reflect the strength of opposition from outside interest groups. The gamut of public ownership outcomes ranges from full nationalization of the oil industry without concessions to multinational groups, as in Mexico, to the absence of state-owned enterprise because of concessions to both multinational and domestic groups, as in the United States.

In the second part of this chapter I analyze the extent to which two groups of countries display these concession patterns. The first group—Mexico, Saudi Arabia, Iran, and Italy—display, I argue, variations on the multinational concession pattern. The second group—the United States, Britain, Japan, and France—shows how limited state entrepreneurial outcomes are when both multinational and domestic concession patterns are present. Why did the United States try only once and fail to create a state-owned oil company? And why was there such a range of state control of oil across the eleven countries examined? Finally, I sum up the thesis of this book: that the relative political opposition of multinational and private domestic groups to state autonomy in oil explains the varying degrees of government equity participation in national oil industries.

DOMESTIC OR MULTINATIONAL CONCESSIONS?

The cases of Norway and Britain reveal a pattern of domestic concessions very different from the pattern of multinational concessions to be found in Indonesia and Malaysia. The domestic concession pattern occurred in Norway and Britain because domestic industrial, labor, and environmental groups were strong opponents of state oil policies. It was characterized by a high degree of political organization in shipping associations, fishing organizations, and trade unions. In the two less developed countries, by contrast, such organizations were either weak or nonexistent.

Second, the domestic concession pattern depended upon domestic groups controlling pivotal, or swing, seats or votes in parliamentary elections or policy decisions. Such elections and decisions threatened to jeopardize the governing coalition's control of the government or its pursuit of expansionary oil policies. Fishing groups held this kind of pivotal power in both Norway and Britain; so did shipping groups in the former country and labor groups in the latter. In neither of the two less developed countries did domestic groups hold such pivotal power. Indonesia's government party (Golkar) held a large parliamentary majority, as did the National Front party in Malaysia after 1973.

Finally, bureaucratic agencies were influenced by domestic group interests in the two industrialized countries much more than in the two less developed countries. Ministries of trade and industry in Norway and Britain were much more susceptible to pressures from powerful domestic shipping associations than were their counterparts in Indonesia and Malaysia. In both LDCs shipping was either wholly or partly government-owned, and pressure from private interests in Malaysia's national shipping company had little impact on the actual investment strategy that the company followed in the 1970s. Military groups, not private industrial interests, composed governing elites and had by far the strongest domestic influence over government oil policies.[1]

But the relative absence of domestic concessions in the two less developed countries does not mean that they were entirely free of societal opposition. Instead, the oil policy process in both Indonesia and Malaysia was dominated by concessions to multinational groups. This multinational pattern reflected the degree to which multinational companies had penetrated national oil-related markets. It also reflected the degree to which governments relied on tax revenues from production by the multinationals. Both Indonesia's and Malaysia's oil production and tax base had been dominated by the Majors or their subsidiaries since the early 1900s. Only in the 1970s did state oil companies begin to carve out a share of national oil production for themselves. In contrast neither Norway nor Britain even started national oil production until the early 1970s, and shortly afterward they began to create both a market share and a profit share for state oil companies.

The strength of domestic or multinational opposition could have altered either the government's share of state oil company ownership or the degree of vertical integration of the company in national and international oil industry operations. In the domestic concession pattern, despite domestic opposition, state oil companies remained wholly or partly owned by the government. Domestic opposition, however, limited the vertical integration of state oil operations within the national industry. In Norway, for instance, state ownership was restricted to a share of production, pipeline, and refining capacity. Concessions to private domestic groups included shares of production, national and international transportation, and refining capacity and contracts. International transportation was primarily left to the multinationals, however, as was a share of production and the larger part of oil exportation and marketing.

The multinational concession pattern resulted in much greater gains for the state. Both Indonesia and Malaysia retained full government ownership of their state oil companies. Would their ownership have been diluted had there been stronger opposition from multinational groups? Or did the prevailing pattern of 100 percent state ownership of oil companies in OPEC countries set a precedent for most developing countries? The greater gains for less developed countries showed up in the vertical integration of their state oil companies. Because of weakness of domestic opposition, the Indonesian and Malaysian governments had to make concessions only to multinational companies. Indonesia gained shares of production and overseas marketing (notably in Japan) and monopolies over national tanker transportation and subcontracting. Malaysia gained a share of production and a monopoly over international shipments of liquefied natural gas. Concessions to multinationals primarily concerned international operations. The MNCs retained their monopoly over international oil transportation and their shares of production, subcontracting (in Malaysia), refining, and marketing.

The main difference between the two concession patterns as they appear in the four countries we have examined is in focus, whether on national or on international operations. The domestic concession pattern gave private domestic groups a share of national operations; the multinational pattern retained a share of international operations for multinational companies.

These two patterns verify the central hypothesis presented in Chapter 2. The two LDC states, Indonesia and Malaysia, gained larger shares of corporate ownership and a more extensive vertical integration of oil operations because they faced less opposition from domestic groups than the governments in the two industrialized countries, Norway and Britain. Can we expect to find the same distinction between other advanced industrialized and less developed countries?

THE MULTINATIONAL CONCESSION PATTERN

The cases of Mexico, Iran, Saudi Arabia, and Italy conform to the multinational concession pattern. Multinational oil companies generally opposed the expansion of state-owned oil companies, independent of mobilization by domestic groups. The cases also demonstrate, however, that even strong opposition from multinational corporations and sanctions by foreign governments were insufficient to dissuade these primarily LDC governments from expanding public ownership over oil for nationalistic purposes.

Mexico

In Mexico nationalization of the oil industry in 1938 was the state's response to unsuccessful bargaining with foreign oil companies. The government began during the early 1920s to assert more national control over petroleum exploitation. In 1921 it tried to double the tax on oil exports to gain a share of the 40 to 60 percent profits accruing to such foreign oil companies as Shell and Standard Oil of New Jersey. But these multinational corporations responded swiftly and deftly. They halted production, putting twenty thousand workers out of work, and they shifted production operations to Venezuela. To stop the evacuation, the Mexican state had to withdraw its oil export tax, and even agree informally to consider compensating the multinationals for losses they had incurred earlier, during the Revolution.[2]

But this setback did not weaken the Mexican state's determination to gain a share of control over national oil. In 1925 Mexico passed the Petroleum Law, which required foreign oil companies to apply for "confirmatory concessions" on their pre-1917 holdings. The multinationals were furious. Backed by the U.S. ambassador to Mexico and the U.S. secretary of state, the corporations challenged the law before the Mexican Supreme Court and even campaigned for a military confrontation between U.S. and Mexican authorities. But New York bankers, realizing the importance of Mexican oil investments, persuaded the Majors neither to break relations with Mexico nor to use force. The issue was resolved in 1927 when the Mexican Supreme Court declared the Petroleum Law unconstitutional; new legislation of a "softer" nature was passed.

Again the Mexican state was not deterred from its aim. It formed a 100 percent state-owned oil company, Petróleos de Mexico (PEMEX), in 1934. A year later the first oil industry union, the Union of Mexican Petroleum Workers, was also created. The coalition that formed between state and union eventually led in 1938 to the nationalization of the Mexican oil industry.

The confrontation with the foreign companies began over wage and benefit contracts for petroleum workers. Foreign companies failed to meet the deadline set by the Federal Labor Board with the approval of the Mexican Supreme Court. President Lázaro Cárdenas then enrolled the support of the secretary of communication and public works and expropriated the foreign holdings of the sixteen U.S. and British multinational oil companies that controlled 98 percent of the national oil industry. This action defied the wishes of the secretary of finance, who urged restraint to protect Mexican loan negotiations with the United States.

The British government broke diplomatic relations and instituted an oil boycott, but the U.S. government used more restraint. It applied diplomatic pressure for oil company compensation and canceled silver purchases from Mexico. Despite British and American protests, however, the Mexican government did not denationalize the Mexican oil industry. It did incur tremendous financial problems, including a devaluation of the peso and a decline in export earnings, which resulted from the loss of silver exports and the oil boycott, respectively.

Nora Hamilton argues that the Cárdenas government was able to nationalize foreign oil investment because of two main reasons. First, the world depression and the U.S. government's shift from oil to manufacturing interests in Latin America limited the extent to which multinational oil companies could gain U.S. government support to oppose the actions of the Mexican government. U.S. and British government restraint about using force to oppose the nationalization allowed the Mexican state to be relatively autonomous. Second, the state had formed a powerful alliance with the working class which propelled the nationalization despite concern from domestic landowning elites. The Cárdenas regime had already opposed the landowners when it enacted extensive land and agricultural reforms benefiting Mexican peasants. Now PEMEX unions were to be paid off in an open-ended bargain.

But the Mexican state had to pay for the autonomy to create PEMEX and nationalize the oil industry with concessions to the multinational corporations. In 1942 an agreement was reached with the Majors and the U.S. State Department on compensation and indemnification. President Roosevelt had insisted on "fair indemnity"; the Mexican debt to the United States was $24 million, of which $18 million was for oil properties. The Mexican government agreed to compensate the foreign oil companies for more than just the value of their oil properties in Mexico. At the same time, because of its need for oil supplies during World War II, the U.S. government also lifted its boycott of Mexican petroleum. Nevertheless, by the end of the war 90 percent of Mexican oil was still being consumed in domestic rather than international markets.

The Cárdenas government also had to give concessions to private domestic groups to build support for the nationalization. These were, however, secondary to the concessions made to the multinationals. They included reductions in the amount of land distributed under land reforms and in the number of peasants benefiting from agricultural reforms.

The Mexican case is one variation on the multinational concession pattern. A multinational oil group including several of the Majors, the U.S. secretary of state, and the British Foreign Office strongly opposed the Cárdenas government's creation of PEMEX and the nationalization of Mexican oil. However, the group did not use military force and merely boycotted oil imports and halted silver purchases. This decision favored the Mexican state's nationalization, but it also softened Mexico's bargaining position, and the state eventually compensated the multinationals for their oil properties. But the Mexican case also diverges slightly from the pattern in that an alliance was built between the Cárdenas government, the labor unions, and (to a lesser degree) private domestic landowners and agriculturalists. The domestic concessions involved, however, were clearly subordinate to much more important multinational concessions.

The Mexican nationalization involved a twofold transfer of wealth. The transfer of oil properties from foreign oil companies to the Mexican state satisfied the national interest in indigenous control of oil. But PEMEX's decision to pay off the unions was a political move designed to reinforce domestic support for the nationalization and for public control of oil. Domestic politics, therefore, were not quiescent but strongly favored public control by the state, through which unions might also retain influence.[3]

Iran

The Iranian government had already begun leasing national production rights to the Majors by the turn of the century. By 1914 private British oil interests had created and established ownership in the Anglo-Persian Oil Company (APOC), eventually to become British Petroleum. Soon after the British government bought up 52 percent of APOC stocks, a controlling share that ensured the Royal Navy of an oil supply independent of purchases from foreign suppliers. These early private and government moves established British control of the emerging Iranian oil industry.[4]

Foreign dominance kindled the same nationalistic fire in Iran as in Mexico. Although the Iranian government received substantial royalty payments, Iranians did not benefit much from the development of the oil industry. British oil profits were deposited in foreign banks or reinvested elsewhere. The Iranian oil sector employed less than 1 percent of the national labor force. Finally, foreign oil operations had no multiplier effect to stimulate growth in other Iranian industries.

By 1949 the Iranian government had become dissatisfied with its royalty payment share. To make matters worse, supplies for the domestic oil market were insufficient, and the British Navy was buying Iranian oil at lower prices than Iranians were paying. The Iranian government decided to nationalize the oil industry in 1951 and create a fully state-owned oil company, the National Iranian Oil Company (NIOC), to control operations. The move put the Iranian and British governments on a collision course. The British refused to buy Iranian oil even though the British Navy was still dependent upon that previously British-controlled source. Like the Mexicans, the Iranians did not budge from their insistence on nationalization. But the loss of British expertise and oil marketing left Iranian production virtually at zero between 1951 and 1954.

Refusal of the British and American governments to negotiate compensation regarding Iranian oil led to the overthrow of Premier Muhammad Mossadegh in 1953. By 1954 the Iranian government was forced to cooperate with the international oil companies in order to increase its production and sales. An agreement was signed to create a consortium consisting mostly of Majors and the British Anglo-Iranian Oil Company (earlier APOC, later BP). AIOC was given 40 percent control, and thirteen multinational oil companies (eleven of them based in the United States) bought up the rest of the shares as a form of compensation to AIOC for the nationalization. This group of companies included six of the Majors (Shell, Standard Oil of New Jersey, Standard Oil of California, Texaco, Gulf, and Mobil) and some Independents, such as the French Compagnie Française des Pétroles. The Iranian government retained management control of the industry and collected 12.5 percent in royalties (in cash or crude supplies) and a 50 percent income tax on oil company profits less the amount paid in royalties.

The government was soon dissatisfied with its small operational role in the oil industry. In 1957 it passed the Petroleum Act, which aimed to expand NIOC's operations through planned exploration contracts and diversification throughout the oil industry. This expansion of NIOC was possible because independent oil companies (for example, the Italian state oil company, AGIP) were ready to supply technology and marketing opportunities in exchange for access to Iranian supplies. More important, they were ready to enter joint ventures with NIOC in order to gain access. Three types of joint ventures were possible: a 50–50 split, or a 30–70 ownership split favoring NIOC, or a form of service contract. The first joint venture was signed in 1957 with NIOC, the second with a subsidiary of Standard Oil of Indiana—both Independents and Majors were cooperating with NIOC from the start. By 1971 a total of eight joint-venture contracts had been signed, and by 1981 NIOC's oil exports constituted 20 percent of total oil production.

In 1973 the Iranian government took a final step toward gaining full ownership and management control of the international consortium by abolishing the earlier consortium agreement with the multinational oil companies. The government then established NIOC as operator for all Iranian oil production and gave it full control over the future plans of the consortium. Foreign participants were guaranteed long-term sales contracts for crude in return for relinquishing ownership, management, and operational control within the consortium. NIOC was in full control of the Iranian oil industry by the 1980s.

Like the Mexican case, the Iranian case conforms to the multinational concession pattern. The Iranian government nationalized the oil industry and created NIOC over the strong objections of the British government and private British oil interests, both of which held shares in AIOC. The oil boycott by the British and other Western governments, together with the loss of British expertise, forced the Iranian government to compromise much more than the Mexican government did. The Iranians agreed to an international consortium. But it was actually cooperation with independent oil companies such as Italy's state-owned AGIP that enabled NIOC gradually to gain greater national control and eventually to assume full ownership of the consortium itself. The ousting of the shah, and the shift of governmental control to religious forces under the ayatollah Khomeni, consolidated national control of oil but almost ruined the industry. Even in this final step of state control, however, the government was obliged to concede long-term crude supply contracts to the multinational corporations. Iran thus offers a purer form of the multinational concession pattern than Mexico. Religious groups may have taken control of the government, but Iranian industrialists never played more than a negligible part in the story of their country's oil.

Saudi Arabia

The pattern of multinational concessions in Saudi Arabia resembles that of Iran more than that of Mexico. The first major involvement of multinational oil companies occurred in 1928 when Gulf Oil signed over a concession received the previous year to Standard Oil of California. Socal was the only possible buyer because the Red-Line Agreement, also signed in 1928, prohibited oil companies that were also members of the Turkish Petroleum Company (later Iraq Petroleum Company) from acting independently in the Middle East. The agreement gave Socal an effective monopoly over Saudi production up to 1936.

In 1936 Socal decided to form a new joint-venture company with Texaco called Aramco. The equal-shares venture gave Socal access to Texaco's markets, and Texaco gained access to Socal's Saudi crude. But this widening of American multinational oil company involvement in Saudi Arabia did not stop there. The U.S. government pressured Aramco to sell shares to Esso and Mobil. In 1948 Aramco's shares were divided up, giving 30 percent each to Socal, Texaco, and Esso and 10 percent to Mobil.

The Saudi government remained in a weak bargaining position throughout these dealings. Aramco had a monopoly over Saudi oil production and paid the government almost nothing for the oil. The company was exempted from all government taxes, though it did pay a lump sum upon discovering oil and had to supply the government's oil requirements.

Between 1949 and 1960, however, the Saudi government struck better deals, particularly with Independents anxious to gain access to Saudi crude. In 1949 the government signed a concession agreement with the Pacific Western Oil Company (later Getty Oil) for royalty payments and profit sharing. This agreement pressured Aramco in 1950 to allow the Saudis to collect income taxes on the company's profits less operating expenses, rental, and depreciation. By 1960 the Saudi government had taken a further step toward participation in national oil production by signing a concession agreement with the Arabian Oil Company. This agreement gave the government a direct 10 percent participation in the concession.

In 1962 the Saudi government created a state-owned oil company, Petromin, with the intention of acquiring a larger share of the benefits from Saudi oil for the Saudis. Through oil production the government sought to improve national economic stability and direct revenues to the Treasury. A state oil company would prevent the country from being hurt by the spiraling prices posted by the multinationals. The government also intended to integrate the oil industry, which was isolated from the national economy and from Saudi national development plans. Finally, the government hoped to get many Saudis employed in the oil industry to rectify the situation that had developed under Aramco. The Saudi state hoped to achieve these goals by developing fully integrated national exploration, production, transportation, and marketing for Petromin. But it also intended to form joint-venture or capital-participation agreements with foreign oil companies to expand into other, international phases of the industry.

Petromin's first move was to purchase Aramco's domestic marketing facilities in 1964. The state oil company then entered joint-venture agreements in 1965 and 1967 for exploration and production of Saudi crude. A first joint-venture was 60–40 in favor of the French state oil company, ERAP. The second and third agreements were signed with Phillips and the Sun Oil Group.

These were significant gains for Petromin, but the Saudi government was determined to achieve full control of Aramco. Between 1973 and 1975 the government negotiated with Aramco's owners to increase Petromin's shares. In 1973 the Saudis purchased 25 percent. The following year the government gained a 60 percent interest in Aramco's producing assets. Finally, in 1975, an agreement in principal was reached for Petromin to take over 100 percent of Aramco. The agreement stipulated that the Saudi government would pay a fee for Aramco's owners to continue providing technical, managerial, and operating services. In addition the multinational corporations would get long-term crude lifting contracts. Yearly amounts of crude in these contracts would decrease unless the multinationals entered into new joint ventures with Petromin for exploration and production.

But these forceful moves on the part of Petromin were coupled with a desire to retain the investments of foreign oil companies in Saudi Arabia. For this reason, the government promised a long-term crude supply of 250,000 b/d for any foreign company willing to enter a joint-venture agreement with Petromin for oil-related projects.

Like the Mexican and Iranian cases, the Saudi variation conforms to the multinational concession pattern quite well. The Saudi government was eventually able to create Petromin over the opposition of Aramco. Petromin sought joint ventures with Independents that were willing to cooperate with the state. Eventually, negotiations led to an agreement to transfer full control and ownership of Aramco to Petromin. The concession to the Majors in Aramco—Socal, Texaco, Mobil, and Esso—was long-term agreements for crude supply and an opportunity to expand supply through joint ventures with Petromin. Private Saudi groups, particularly branches of the ibn Saud family, were even less involved in gaining industrial concessions from these oil dealings than were bourgeois groups in Mexico or religious groups in Iran.

In sum, the cases of Saudi Arabia, Iran, and Mexico show that opposition from multinational groups—including foreign oil companies and foreign governments—was not strong enough to deter government intervention in national oil operations. All three governments created 100 percent state-owned oil companies and integrated their investments throughout the national oil economy by nationalizing the entire industry. However, in all cases the state made cooperative arrangements with foreign oil companies to continue access to the overseas oil trade networks controlled by the multinationals.

Italy

The case of Italy differs from those of Saudi Arabia, Iran, and Mexico, in two major ways. First, the Italian state never nationalized the Italian oil industry. Second, however, the Italian state-owned oil company expanded into international rather than merely national oil production despite resistance from the Majors.

Up to the 1920s the Italian market was dominated by foreign multinational oil companies. The Italian state created a wholly state-owned oil company, AGIP, in 1926 for motives that resembled the Mexican, Iranian, and Saudi governments' interests in asserting national control over a foreign-dominated domestic industry. The head of AGIP, Enrico Mattei, aimed to break up the Seven Sisters' monopoly in Italy and establish AGIP as a fully integrated, international oil company. AGIP's problem, however, was that it depended heavily upon these foreign oil companies for oil supplies.[5]

This dilemma lasted until 1953, when Mattei created Ente Italiane Idrocarburi (ENI), a state holding company including AGIP, and declared a monopoly over Italian gas operations. His leadership in ENI allowed the state holding company to dominate government oil policy. Mattei was able to set policy and give financial support to Christian Democratic deputies which helped him influence Parliament. Some opposition came from the Ministry of Industry, which reflected the views of private industry. But Mattei did not have to strongarm government officials, because a large part of the government supported his nationalistic commitment to lessen the dominance of the Majors. Throughout the fifties and sixties Italy's powerful trade unions were particularly strong advocates of state corporate expansion, because public companies offered more jobs and better benefit packages than did private companies.[6]

What private domestic oil companies did exist were so small that they could not resist the state's taking a strong role in oil and gas. Even private industrialist consumers of natural gas submitted to ENI's pricing schemes and efforts to compete in the petrochemical industry. They avoided conflict with ENI to eliminate a much worse possibility: nationalization of the oil and gas industry.[7]

The real contest of wills between ENI and the Majors came when ENI tried to gain a share for AGIP in the new oil consortium being created in Iran. When the Majors rebuffed ENI, the state oil company went behind their backs and in 1957 signed lease agreements with the Iranian government. These agreements shocked the Majors, for they overturned the traditional 50–50 formula for profit sharing established in Venezuela in the 1940s and in Saudi Arabia in 1950. ENI agreed to a 75–25 formula favoring the Iranian government. It also accepted partnership with NIOC to establish oil operations of Iranian nationality. As the Majors were operating agents for the Iranian government, both of these agreements threatened the Majors' control. In 1959, furthermore, ENI made international investments of its own in Austria, Germany, Switzerland, the Sudan, Morocco, and Tunisia.[8]

By the early sixties, however, Mattei was beginning a rapprochement with the Majors. ENI had become a world oil company, and Mattei was ready to cooperate with the Majors on an equal footing. After Mattei's death in 1962 Esso signed a five-year supply agreement with ENI. By then oil discoveries in Libya, Egypt, Nigeria, and Iran had given ENI stature as a multinational oil company; further discoveries in the North Sea later in the decade would give it control over 25 percent of the Italian market by 1971. In 1973 Shell sold its Italian operations to ENI because the Major was dissatisfied with relations with the company and the Italian government.[9]

By the 1980s Italy still depended upon the Majors for oil, and it was second only to Japan in the magnitude of that dependence. ENI was still purchasing 58 percent of its crude from other multinational oil companies, but it had grown to control 33 percent of the Italian market. Furthermore, AGIP was an international oil company with operations in a dozen countries.

Although the Italian government never nationalized the oil industry, it did create a 100 percent state-owned oil company that joined the ranks of international companies. This state entrepreneurial expansion occurred despite the opposition of the Majors, which refused to help Mattei gain Iranian concessions. ENI acquired oil concessions from Iran on its own, however, and eventually bought out Shell's operations in Italy. By the early 1960s the state oil company began to cooperate with the Majors (in particular Esso) to expand its marketing operations overseas. The story of ENI again conforms to the multinational concession pattern in that it features strong initial opposition from multinational groups to state oil enterprise. Domestic groups did not oppose state oil policies. Eventually, concessions from both the state and the multinationals led to a pattern of cooperation for crude supply and marketing.

COMBINED DOMESTIC AND MULTINATIONAL CONCESSION PATTERNS

In the cases of Japan, France, Britain, and the United States both domestic and multinational groups opposed state oil entrepreneurship, and the cases exhibit a combination of domestic and multinational concession patterns. Yet the outcomes for state oil enterprises in the four countries were dramatically different. The Japanese government managed to overcome opposition and created a 100 percent state-owned oil company in 1978. The French and British governments formed 100 percent state oil companies in 1965 and 1976, respectively, but had to release shares of corporate equity to private oil companies. The U.S. government also created a 100 percent state-owned oil company during World War II, but the company survived only six months under a barrage of opposition from domestic and multinational oil companies. How can we explain this range of outcomes when concession patterns were initially similar?

Japan

The Japanese state has been recognized as strong in guiding industrial policy, unlike the U.S. state, which is weak because of its vulnerability to the demands of U.S.-based multinational oil companies. Richard Samuels shows, however, that private industry interests in Japan were consistently able to check the growth of a 100 percent state-owned oil company until 1978, when the Japanese National Oil Corporation (JNOC) was established. As a result of this persistent opposition, the state company managed to develop operations only in stockpiling, exploration (fields had to be divested upon discovery) and production using surplus foreign currency from a positive balance of payments. This limited outcome for state entrepreneurship was less than the original intentions of the Ministry of International Trade and Industry (MITI).[10]

State intervention in the Japanese oil industry was from the start shaped by political conflict among bureaucratic, private domestic, and multinational industry interests. During the 1920s the Japanese industry was dominated by four private domestic companies, all nationally-owned: Nippon Oil, Kotura Oil, Mitsui Oil, and Mitsubishi Oil. But by the mid-twenties the American company Standard Oil and the British Rising Sun Oil had carved a niche in the industry for foreign oil companies by supplying 70 percent of the crude for the Japanese market. Japanese companies acted as distributors.

Although the Japanese government tried to increase its role in oil, strong private domestic oil interests kept such efforts at a minimum. The 1934 Petroleum Industry Law was one such attempt. It empowered the Ministry of Commerce and Industry to require private companies to stockpile oil. It also enabled the government to license, allocate market shares, and fix prices in addition to providing tax assistance, financing, and protective tariffs to private domestic oil companies. But the major Japanese oil refining groups—there were eight of them by then—were still strong enough to prevent the state from gaining production control within the industry. Even the 1938 Petroleum Resources Development Law enabled the state only to subsidize private domestic exploration as a hedge against embargoes by the Americans, British, or Dutch during the war with China. In exchange the state got a 2 percent royalty on the oil produced. Although the state made several efforts between 1935 and 1940 to coordinate investment by private Japanese oil companies and subsidize exploration and distribution within Japan, none of these efforts led the state to take an entrepreneurial role in production.

State intervention in oil peaked during World War II, when the government nationalized the oil industry and created a partly state-owned oil company. Imperial Oil Company was created in 1941 following a cut-off of U.S., British, and Dutch petroleum supplies.

The Ministry of Finance provided an initial $280,000 and the rest of the company's shares were bought by private Japanese companies; Nippon Oil was the largest private shareholder with 61 percent of private stock. Imperial Oil invested in operations in Japan, Borneo, and Indonesia, but the state's oil interests were diluted by private interests—the head of the company was formerly president of Nippon and the board of directors overrepresented private oil interests.

During the postwar occupation of Japan by the Allies, the Imperial Oil Company was given full responsibility to produce domestic crude (of which it already produced 90 percent). But the company was also directed to decrease the concentration of private power in the company by increasing the number of private shareholders.

After first supporting government involvement in oil, however, the supreme commander for Allied power decided in 1950 that Imperial Oil was ready to be fully privatized. The Majors were resuming exports to the Japanese market, and the company no longer needed government participation or the privileges normally granted to public companies. Once its public shares had been sold, Imperial Oil was allowed to move into transportation, refining, and marketing, from which its public charter had excluded it.

The Japanese state did not make another major attempt to set up a state-owned oil company until 1955. Then it established the Petroleum Resources Development Company (PRDC), forerunner of the current Japanese National Oil Company (JNOC). The PRDC was modeled after Imperial Oil, and the state was given two-thirds ownership. The purpose of the company was to reduce Japan's dependence on foreign oil supplies controlled by the multinationals and on foreign capital. Both dependencies had resulted from long-term supply contracts negotiated during the occupation and upon financial loan agreements with the Majors.

The creation of PRDC did not occur without internal bureaucratic conflict. MITI insisted on the formation of the company, but the idea was strongly opposed by the Ministry of Finance, which was concerned about the extent of public capitalization required. Nor was the company created without resistance from the private Japanese owners of Imperial Oil and other, foreign oil competitors. Mobil, Standard Oil of California, Standard Oil of New

Jersey, Texaco, and Royal Dutch Shell all owned substantial shares of the main Japanese oil companies, suggesting that foreign oil interests played a large part even in "domestic" industrial opposition to state enterprise in oil. Bureaucratic, domestic, and multinational resistance was overcome when the government agreed not to guarantee the PRDC's liabilities. Private oil firms would have equity participation in the company, and the company would assist private firms in new exploration.[11]

But the 1973 global oil crisis brought home Japan's need to reduce its dependence on foreign oil companies for 70 percent of the country's domestic supplies. The government needed overseas sources of its own. Its main tool, the PRDC, had been upgraded and renamed the Japan Petroleum Development Corporation in 1967 after MITI agreed only to finance private overseas oil projects and to divest its interest once exploration had been completed, a concession to domestic and multinational oil interests (particularly the Getty-affiliated Mitsubishi Petroleum Company). But by the oil crisis years of the early seventies MITI was again pushing for a state-owned international oil company. In 1978 Japan Petroleum was renamed the Japan National Oil Corporation (JNOC). MITI's Advisory Committee for Energy insisted that for national security reasons, the new company should ensure that domestically controlled supply would service 20–30 percent of Japanese oil needs. JNOC was also made a 100 percent state-owned company.[12]

The company remained primarily a financing agent for domestic and overseas exploration by private Japanese oil firms. It can extend loans for 70 percent of an overseas project or for more if the principal operator is a Japanese firm. JNOC can also offer direct financial support to foreign governments interested in expanding their exploration and production projects. But the Japanese state company, unlike Italy's AGIP or Britain's Britoil, is not empowered to run its own oil exploration or production.

The Japanese case demonstrates the persistence of a state intent upon achieving government control and ownership of oil despite strong resistance by Japanese and multinational oil companies within the domestic industry. The relatively strong and autonomous MITI was able to orchestrate an overseas oil policy in which the 100 percent state-owned JNOC eventually took part. MITI insisted that the Japanese oil market should not be entirely dominated by the multinationals or by Japanese oil companies in which those foreign companies held major shares. However, this entrepreneurial state oil policy also required the cooperation of domestic Japanese oil companies and the Majors. The concession that these companies received was that once JNOC discovered overseas resources, actual production would fall to private Japanese oil companies—either Independents or companies in which the Majors held a share.

France

The French case, like the Japanese, shows state perseverence to participate in the national oil industry. But it also demonstrates how strong multinational and private domestic oil groups forced concessions on the state. The difference between the two cases is that the French state oil company was partly privatized. Conservative leadership under Giscard d'Estaing in 1974 weakened the state's role in oil by creating a limited partnership with private oil interests.

By the 1920s the Majors dominated the French oil market. Esso controlled 56 percent and Shell 15 percent of all oil supplies; together they controlled over two-thirds of the French market. This foreign domination created strong nationalistic feelings regarding French control of oil. French marketers encountered difficulty, and in response the French government passed a petroleum law in 1928 which aimed to secure domestic supplies and develop the national oil industry.[13]

The contest between the French government and private domestic oil produced a decision favoring state control of the French share of the oil industry's sales and refining. The modus operandi set up in 1928 molded the French industry for years. The government gained a 35 percent share in the Compagnie Française des Pétroles (CFP), originally established in 1924, with 35 percent voting control, and a 10 percent share in CFP's refining subsidiary, despite the opposition of the French distribution companies that held shares in the company. French industrialists had originally created CFP in 1923 to participate in the Turkish Petroleum Company, which later became the Iraq Petroleum Company (IPC) producing oil in Iraq for the French market. Other shareholders in the Turkish company were several Majors: Anglo Persian (later BP), Standard Oil of New Jersey, and Standard Oil of New York. What started as a cooperative, nationalistic effort to develop French oil capacity thus ended in a reluctant partnership between the state, French industrialists, and the Majors. State enterprise was made possible, as Harvey Feigenbaum points out, by the conflict of interests between private domestic and international oil companies.

The government attempted to make this cooperation more palpable by establishing a "delegated monopoly." It directed oil industry development and then allowed private companies to organize operations themselves. Although the government's role did not involve wholly state-owned enterprise, it did enable the state to allocate market shares to oil companies as the overall industry expanded and to determine suitable returns on investment. In 1939 the government increased its control over exploration and production by establishing the Régie Autonome de Pétrole (RAP), in which the state gained a 51 percent share. But a new company, Société Nationale des Pétroles d'Aquitaine (SNPA), was also created in 1939, and although jointly owned this second company represented private more than public interests in exploration, production, and distribution.

The postwar government under Charles de Gaulle strengthened the French state's control of the oil industry by creating a 100 percent state-owned oil company in 1945. This company, the Bureau de Recherches de Pétrole (BRP), was to reduce France's dependence on foreign oil companies for oil supplies. The state oil company was to conduct exploration and production in collaboration with private oil companies that had majority French holdings. But the new state-owned company also cooperated with the Majors and Independents in its expansion overseas. By 1950 the company had invested in exploration in France, Equatorial Africa, and Algeria. These investments were made jointly with private multinationals, including Mobil, Shell, and American and European Independents.

These state oil company forays into the French and overseas markets provoked apprehension from the Majors. Esso, Shell, BP, and Mobil held dominant shares in distribution for the French market. They also had large investments in crude production in Algeria and Equatorial Africa, where BRP was now expanding and offering management contracts (*contrats d'association*) to host governments more favorable than the Majors' concessionary agreements.

In 1965 the French government attempted to rationalize the state's role by consolidating BRP (100 percent public), RAP (51 percent public), SNPA (public- and private-owned), and a new public-private company, UGP. The newly created national oil company, ERAP, was 100 percent state-owned and included both holding and operating companies previously controlled by BRP. ERAP marketed petroleum products under the trademark Elf (Essences et Lubrifiants de France). However, SNPA continued to be run separately because of its private shareholdings. Private French companies wanted to keep the state at bay, particularly as they had already given up the struggle to compete with the Majors. They had instead joined with the Majors to form Total, which was then taken over by the jointly French and foreign-owned CFP. By the end of the 1960s French companies—ERAP (100 percent public) and SNPA and CFP (both public-private)—controlled 23 percent of the French market.

But these moves toward a 100 percent state-owned French national oil company were eroded under the conservative government of President Giscard d'Estaing, a former finance minister. In 1976 ERAP and SNPA were merged to form a new, vertically integrated oil company, Société Nationale Elf-Aquitaine (SNEA). The French government had an equity share of 70 percent in SNEA but controlled only 52 percent of its corporate operations. ERAP became the 100 percent state-owned holding company for the state's 70 percent share in SNEA. The French Left claimed that the conservatives had denationalized ERAP; multinational and French domestic oil interests saw the merger as providing a way for private international capital to enter SNEA (the 30 percent private share) without a stifling degree of government control. By 1980, however, SNEA had failed to increase the French share of the domestic market to above 23 percent.[14]

The French and Japanese cases demonstrate the combined effects of multinational and domestic concession patterns when both multinational and domestic groups oppose state entrepreneurship in oil. The French outcome, 70 percent state-ownership of SNEA, represents more of a compromise than does the Japanese case, 100 percent state ownership of JNOC, regarding state equity participation in the state oil company. The French state was eventually able to overcome the opposition of powerful multinational and domestic oil interests, and in 1945 it achieved a fully state-owned oil company. Weakened by an incoming conservative coalition in the 1970s, however, the state conceded a 30 percent share of its control to private oil interests.

I agree with Harvey Feigenbaum that this picture of compromise challenges conventional notions, derived from other kinds of industrial policy, of the French state as strong and autonomous. Feigenbaum explains "lack of control" by the state as a result of state elites becoming ideological captives of the private sector, just as the U.S. bureaucracy is captured by narrow interest groups in Grant McConnell's description.[15] I argue, rather, that the French bureaucracy compromised public oil enterprise for structural reasons. The state could not have sustained its role in oil without making concessions to private domestic and multinational oil groups that eroded both the state's equity share in SNEA and SNEA's French share of the domestic market.

The French case is similar to British case in that concessions eventually led to a reduction of the state's shareholding in the state oil company. It is different in that the British state was even weaker in pursuing autonomous state oil interests than its French and Japanese counterparts.

Britain

British investment in overseas oil resources started in the Persian Gulf as early as 1901. In 1909 this initial investment by private British interests became the Anglo-Persian Oil Company (APOC)— later to become one of the Majors as British Petroleum.[16]

The British government first played a role as an oil entrepreneur soon after. In 1914 Winston Churchill, the first lord of the Admiralty, persuaded the British government to acquire a controlling share in APOC. The Royal Navy needed a secure supply of oil, and Churchill feared that the navy's former suppliers, Standard Oil and Royal Dutch Shell, would overcharge. The British government bought up 52.5 percent (later falling to 51 percent) of the company's shares. After 1914 British private and state interests would be involved in a joint venture to produce petroleum overseas.[17]

Between 1914 and 1954, however, relations between the government and private shareholders in BP became increasingly strained. Succeeding governing coalitions pressured BP and Shell (40 percent privately British-owned) to promote British interests and thereby to offset growing American dominance in the oil industry. Although the government controlled two seats on BP's board of directors and could veto decisions, it refrained from exercising its power except on questions touching on foreign and military policy. But strains between the actual and the nominal influence that the British government could exert over BP policies came to a head during the Iranian nationalization of 1951–54. The British government realized then that its naval interests in oil supplies for Britain were only "roughly parallel" to those of private British shareholders in BP.[18]

By the end of the 1973 oil crisis the British government's relations with BP had reached a turning point. BP had continued to honor foreign contracts rather than divert oil supplies to Britain as the government had wished. After examining the issue of BP control, the Foreign Office became convinced that British government interference in BP's relations, with the U.S. government, for instance, would severely jeopardize the company's future interests in oil areas such as Alaska.

This stalemate over BP control led the new Labour government in 1976 to create a fully state-owned oil company, BNOC. The government aimed the new company's efforts at oil resources in the British sector of the North Sea rather than at overseas opportunities where BNOC might compete with BP's interests. By 1970, 65 percent of the area open for bids in the British sector had been licensed, primarily to such Majors as BP, Shell, and Esso. But state oil companies were involved in producing domestic oil supplies in so many other countries as to give a certain legitimacy to the concept in Britain as well. BNOC's primary directive, therefore, was to provide a reliable supply of government-controlled oil, starting with national production but eventually expanding into international operations.

The dominance of the Majors in British industry deterred BNOC's transformation into a fully integrated international oil company. BP, Shell, and Exxon continued to play a major role in Britain's national oil production. These Majors also continued to own about two-thirds of the U.K.-registered tanker fleet, which gave them an ability both as oil producers and as consumers and producers of tanker services to influence the direction of BNOC's investments.

What is surprising about the British story is that even the incoming Conservative government under Margaret Thatcher could not get rid of BNOC, renamed Britoil. By the mid-1980s the company's profitability and total assets made its production operations too valuable to divest, although the government did sell its trading arms and reduced its 100 percent ownership of Britoil to a minority shareholding of 49 percent and its share in BP to 31 percent.

The British case, like the Japanese and French cases, demonstrates the ability of persistent states to retain government equity control of oil operations. The main difference is that the British state ended up with only minority state shareholdings in both Britoil and BP, while the Japanese and French states retained 100 percent and 70 percent equity control, respectively, of their state oil companies. As a public manager of national oil industrialization, the British government under Thatcher was weaker than the other two states. Thatcher attempted to privatize Britoil entirely by selling off public shares to the multinational oil companies. The government was prevented from doing so, however, because the impoverished British Treasury could not wean itself away from the high profits and long-term foreign exchange earnings of the state oil company.

In contrast to Britain, France, and Japan, the United States is a case of the absence of the sovereign oil entrepreneur. U.S.-based multinationals combined forces with domestic oil companies to dash the state's feeble attempt to create an oil company owned by the American state.

The United States

The United States is particularly interesting because it shows a government trying but failing to institute state enterprise as a major part of its oil industrial policy. Although the particular entrepreneurial effort discussed here occurred during World War II, it is important that we understand it, because it is the only time in U.S. history that a state-owned oil enterprise came close to becoming an American reality. The case is also important because it adds a new dimension to the analysis offered in this book: some governments never succeed in using entrepreneurship as an oil industrial policy. They fail because they lack the state strength and the persistence to overcome multinational and domestic opposition. Whether the U.S. government should be actively involved in the international production and distribution of oil was an issue on which U.S.-based multinationals and domestic Independents had the final say.

World War II forced the U.S. government to consider producing and distributing strategic oil supplies. The war showed that the country was vulnerable to supply shortages and could not rely entirely on its own reserves in the future. By 1941 net exports of oil were only 0.8 percent of total consumption, compared to a range between 3.6 percent and 12.3 percent in the previous three years. At the same time U.S.-based multinational oil companies were discovering potentially massive new reserves in the Middle East which, if the United States did not act quickly, would fall into British hands.[19]

In facing this difficult situation with the British, involving natural resources, trade, and diplomacy, the U.S. government also saw that it could serve its own interests in managing oil supplies. Those national interests were independent of the corporate profit and market-share interests of private, U.S.-based, multinational and domestic oil companies. At some time during 1943 the Roosevelt administration, the departments of State and Interior, and the navy all considered expanding their control of oil supplies in a concerted government effort. But they were not concerned about domestic supplies; rather, the U.S. government was worried that the British might gain control over the newly found reserves in the Middle East.

The Roosevelt administration wanted to produce oil—even if it was foreign oil—autonomously for reasons similar to the early oil interests of the British government. Industrial development could hardly have been the objective, because the U.S. oil industry already had multinational companies producing abroad as well as Independents producing in Texas, Louisiana, Oklahoma, and Wyoming. Even during the war U.S. nationalism was not directed toward increasing government control over oil supplies. But as in Britain, the government needed to increase its control over oil supplies for national security reasons. The threat of oil shortages for the remainder of the war and in the postwar period spurred U.S. leaders to seek direct government control and to reach beyond national oil-producing areas to foreign ones. For similar reasons, the government also wanted to prevent the United States from becoming a net importer of oil. The new reserves in Saudi Arabia thus became a prime target for U.S. government control.

Although there is some disagreement about whose initiative came first, representatives of both the U.S. government and the U.S. oil industry began to explore options for expanding government control over Saudi oil. There seemed to be three options: create a state-owned oil company to gain a share of Saudi production: offer U.S. aid to the Saudis in exchange for assured oil supplies; or give diplomatic support to private U.S. oil companies already working in the peninsula in exchange for a guaranteed U.S. oil supply. Socal (Standard Oil of California) and Texaco (then the Texas Company) already jointly owned Aramco (then Casoc), which had been working in Saudi Arabia since the 1930s. However, exclusive reliance on Aramco to provide and secure the supply of oil from Saudi Arabia, especially during wartime seemed risky to government leaders. Besides, the British had their eye on new Saudi oil reserves and, unless the United States moved quickly, would attain a dominant position to control these supplies.

U.S. government leaders therefore turned to the idea of a state-owned oil company as the best way to increase direct government control over U.S. oil supplies. According to Irvine Anderson, the idea of a state oil company first emerged in the Department of State's Committee on International Petroleum Policy. The committee was particularly interested in gaining control over foreign supplies as a way to slow the drain on Western Hemisphere strategic resources and extend U.S. aid to Saudi Arabia. But, it was actually the secretary of the interior and petroleum administrator for war, Harold Ickes, who adopted the idea in 1943. Pushing aside the State Department's suggestion that the state company acquire option contracts on Arabian oil, Ickes proposed instead that a 100 percent government-owned oil company, the Petroleum Reserves Corporation (PRC), be set up to buy out Socal and Texaco's entire control. The two companies would be reimbursed for all expenses and granted a percentage of future PRC production. Once proposed, the idea attracted support from Roosevelt and the navy. Both were concerned, along with the departments of Interior and State, about increasing government control over foreign oil supplies for strategic purposes. What is somewhat surprising, however, is that Socal and Texaco seem to have welcomed direct government participation in Saudi production. Ultimately they were willing to give only one-third of their control over Aramco to the PRC. In return for this share they wanted government military and diplomatic protection in the Middle East, legal protection from antitrust suits on their operations, and federal financing for their refinery in the Middle East.

The PRC was officially established in 1943 as the government's fully state-owned oil company, with Ickes as its president. The Board of Directors consisted of the secretaries of state, interior (Ickes), war, and navy. State was the only department to gain a veto power and the right to negotiate all agreements with foreign governments. No private oil industry groups were included in the state company's leadership or governing board—the omission reflected the audacity of the wartime government. The PRC had the right to manage the development of all Middle Eastern oil reserves for the United States, but Socal and Texaco were to get preference when bidding against other oil companies for production operations.

Opposition to the state-owned oil company came from multinational oil companies not included in the deal. Small Independents from the domestic U.S. industry opposed the PRC because it threatened to flood the U.S. oil market they controlled with Middle Eastern oil. U.S.-based multinational oil companies, the Majors, were also opposed (except obviously for Socal and Texaco) because control of Middle Eastern reserves by a state oil company was directly competitive with their own interests in those reserves. If the government edged out private companies in the bid for new oil concessions abroad, it threatened to set a dangerous precedent. The Majors had a strong voice through their representation on the Foreign Operations Committee of the Petroleum Administration for War. Standard Oil of New Jersey (now Exxon) and Socony-Vacuum (Mobil) may have been particularly opposed to the Saudi deal because they were interested in a share of Aramco for themselves. They could not get shares if the government took a share too.

But Socal and Texaco were not willing to give private competitors a share of Aramco operations. By excluding Exxon and Mobil, Socal and Texaco could develop Saudi reserves exclusively and thus quickly outdistance their competitors in operations and profit. So two Majors, Socal and Texaco, sided with the government in its attempt to become an oil entrepreneur, but the rest of the Majors and U.S. Independents opposed state oil enterprise. They launched their attack from the Congress.

A coincidence of pressure from Major and Independent companies as well as direct pressure from Congress and the state of Texas, on the departments of State and Interior eventually led to the PRC's downfall. Anderson argues that the crucial blow came once negotiations between the PRC, Socal, and Texaco began to close on a one-third share for the government company in exchange for government financing of the Ras Tanura refinery. Using Ickes's confidential diary, original PRC records, and unpublished State Department documents, Anderson suggests that Ickes abruptly terminated the discussions on 15 October 1943, after a meeting with the president of Mobil (John Brown). Brown apparently convinced Ickes that direct government participation in the oil business would disrupt existing world oil arrangements, provide no great improvements in the government's existing ability to use private oil reserves in times of emergency, and contribute to political disruption in the United States. The result of this meeting, and the subsequent withdrawal of support by the Department of State, was the demise of the PRC in 1944.

Before the company ultimately fell, however, Ickes had made great strides. The PRC was involved in refinery and oil pipeline plans that would be the basis for a vertically integrated oil company with investments in production, refining, transportation (via pipeline), and marketing in the United States. Exxon and Mobil, backed by the Department of State and the British, blocked the PRC's stock-purchase plan for the refinery. In 1944 the PRC entered an agreement with Socal, Texaco, and Gulf to build a pipeline from the Persian Gulf to the Mediterranean with government money. Apparently Ickes made the agreement in secret. Once Congress and the Independents found out, they gave the PRC its death blow. Independents based in Texas and others not involved in the Middle East feared that worldwide regulation of production and imports to the U.S. would ruin their control over the domestic market. They pressured Congress to dissolve the PRC to prevent the emergence of a government-owned, controlled, and organized oil sector that would compete directly with private companies.

With its entrepreneurial oil policy a failure, the government turned in 1944 to a previous policy option: it formed an international agreement with the British to secure a U.S. share of Middle Eastern oil. The agreement in effect proposed an international cartel to manage Middle East production. But even this option met with eventual defeat because U.S. domestic Independents feared the flood of cheap foreign oil into the U.S. market. Congressional opposition and nonratification by the Senate in 1952 sounded the cartel's death knell.

In the end an idea originally in disfavor, because it did not give the government sufficient control over strategic supplies, was chosen. It would mold industrial policy in the United States for years. The oil industry, oil supplies, and oil operations were simply left in the hands of private multinational oil companies. The government also backed Exxon and Mobil in their demand for a share of Aramco, against the wishes of Socal and Texaco. The latter two companies may in the final analysis have been willing to open Aramco to Exxon and Mobil to avoid the added risks if exclusive Saudi investment meant a full-scale marketing battle with Exxon and Mobil. Broader-based ownership of Aramco would also stabilize world oil markets and secure the financing needed to replace the share of the pipeline financing which the government had previously agreed to shoulder.

So prevailed the liberal notion of government as a sovereign but not an entrepreneur. Powerful private domestic and multinational oil company interests prevented the implementation of autonomous state interests in state oil entrepreneurship. The state retreated to a policy of merely taxing and regulating the oil operations of private companies overseas and on its homelands.

The inability of the U.S. state to sustain a state-owned oil company is surprising when we consider that the Japanese, French, and British governments were able to do so even though powerful multinational and domestic oil interests dominated their oil industries also. The British case most nearly resembles the U.S. case. The key difference is that the British state was more insistent than its American counterpart about government control of oil. BP and Shell were both partly state-owned and therefore more willing to avoid conflict with the British state. In the U.S. case, by contrast, multinational and domestic opposition to the state oil company was overwhelming, and state concessions of control were total.

The Japanese, French, British, and U.S. cases demonstrate the combined effect of multinational and domestic concession patterns when both multinational and domestic groups oppose state oil entrepreneurship. In the Japanese case the autonomy and persistence of MITI enabled the state to retain 100 percent control of its state oil company and thus a fully government-owned and controlled access to oil production and supplies. It agreed, however, to leave production to private oil companies. The French case is an intermediate outcome. The French state was eventually able to overcome the opposition of powerful oil interests and achieve a fully state-owned oil company. Weakened by an incoming conservative coalition in 1974, however, the state reduced its equity in the company to 70 percent in order to concede a share to private oil interests. In the British case the initially 100 percent control of BNOC seems to have been achieved without great opposition from foreign multinationals, BP, or Shell. But an incoming Conservative government weakened the state's oil position and reduced its shareholding in the state company to only a minority share. The U.S. state was unable to overcome the opposition of international Majors such as Mobil and Exxon and of domestic Independents. It ended up with no state oil company at all. The efforts of the departments of State, Interior, and the Navy to be autonomous were overwhelmed by private societal demands regarding foreign oil policy.

EXPLAINING STATE OIL ENTREPRENEURSHIP IN ELEVEN COUNTRIES

At some point in the oil histories of all eleven countries considered in this book, national governments formed 100 percent state-owned oil companies to guarantee a national share of oil. The relative strength of private domestic, multinational, and related oil industrial groups was a key factor in determining when autonomous state interests could emerge in the form of state entrepreneurship. Let us group the eleven cases according to degree of political opposition leading to multinational and domestic concession patterns. In Japan, Britain, the United States, and France, as we have seen, governments had to make major concessions of oil operations to both multinational and private domestic groups in order to create state oil companies and insure their survival. Governments in Japan and Britain took longest to erode political opposition to state intervention in oil through a wholly state-owned entity. The U.S. government made its first attempt early in 1943, but was never able to sustain a state-owned oil company because of the powerful opposition of multinational and domestic oil interests.

In six other countries, Italy, Indonesia, Malaysia, Saudi Arabia, Mexico, and Iran, multinational but not domestic groups led the opposition to state oil entrepreneurship. The relatively slow growth of opposition from private industrialists in these countries may explain the relatively greater capacity of governments to create and sustain vertically integrated state oil companies. These governments, moreover, remained highly centralized regarding oil policy, favoring autonomy from domestic interests.

Norway is an exception to both groups of countries. It is the only case in which private domestic opposition to state oil entrepreneurship remained the primary deterrent to the state's expansion. Although opposed to Statoil in principal, the multinationals actually played only a minor role in constraining state oil policy and investment.

These cases support the central hypothesis of this book. The more the eleven governments shifted from implementing private (societal) oil interests to pursuing their own bureaucratic oil interests, the more they relied on wholly state-owned oil companies. Private domestic groups curtailed the expansion of state oil enterprises within national operations in Norway; multinational oil groups constrained the expansion of state oil enterprises in Italy, Indonesia, Malaysia, Saudi Arabia, Mexico, and Iran. Mexico and Iran exemplify the least effective opposition by multinational groups, resulting in only minor concessions—compensation for oil takeovers or long-term supply contracts. Japan, the United States, Britain, and France are particularly interesting in that their state oil enterprises were either slowed in their emergence (Japan), or entirely (the United States) or partially (France, Britain) eliminated once created.

Less developed countries, it seems, gained more as oil entrepreneurs than did advanced industrialized countries. In Indonesia, Malaysia, Saudi Arabia, Mexico, and Iran, state oil companies became fully integrated into exploration, production, transportation, refining, and marketing at the national level and, in most cases, also the international level. Although an advanced industrialized country, Italy has an oil history of vertically integrated state enterprise resembling the pattern common to less developed countries. In contrast, five advanced industrialized countries, the United States, Japan, Britain, Norway, and France, made much less spectacular gains in publicly owned national oil capacity. The United States ended up with no state oil company at all. France ended up with one that was only 70 percent state-owned but that was vertically integrated in international operations. By 1985 the British had a 49 percent and the Norwegians a 100 percent state-owned oil company in national but not international operations. And the Japanese had a 100 percent state-owned oil company with limited stakes in overseas exploration. The greater gains for states that pursued national control of oil profits and supplies through publicly owned oil operations clearly fell into the hands of the LDC group.

What does it mean if state-owned oil companies controlled a large share of global oil trade? And what do greater gains for less developed countries engaged in international oil trade mean for the distribution of future oil supplies?

CHAPTER SIX

Contrasting Perspectives
on the Role of the State

How does the argument in this book square with those of major theories of the state? Some theorists focus on the product cycle, international bargaining, and dependency; others employ bureaucratic and statist interpretations. Some explain neither the emergence of state enterprise nor the behavior of the oil industry, but all address the role of the state. In this chapter I contrast these various explanations of state behavior with my own to identify consensus and controversy in understanding state enterprise in oil.

I examine three aspects of state entrepreneurship: conditions for state enterprise, state institutional interests, and state-society bargaining. First, product-cycle and oil-market theories establish different growth conditions for state entrepreneurship than do theories that emphasize rules and power distribution in the global "oil game." Second, bureaucratic, statist, dependency, and pluralist theories offer different interpretations of state enterprise as a policy of the state acting in its own interests. Third, international bargaining, dependency, and statist models of bargaining between state, multinational, and domestic groups emphasize different constraints on the expansion of state oil companies. Each of these three aspects is crucial to the understanding of state oil enterprise presented in this book.

CONDITIONS FOR STATE OIL ENTREPRENEURSHIP

My argument starts from Raymond Vernon's analysis of product-cycle and oil-market conditions, going on to explain the institutional and political conditions that enabled states to alter oil games at the national level. I thus complement the work of various theorists who explain the reciprocal effects of states on oil rules and distributions of power at the international level.

Product-cycle and market theory explains many of the economic conditions that states must satisfy to enter oil markets as entrepreneurs. Vernon argues that gradually, through the 1960s, governments gained technical skills, knowledge, and experience in the oil industry. At the same time multinational companies began to lose their vertical integration, concentration of world sales, and oligopolistic power in global oil markets. Once states surmounted the barriers to market entry, they were able to refine, market, and eventually produce and transport their own oil. Of significant help to new entrants was OPEC's role in keeping prices high and supplies scarce, thus insuring the windfall profits of the 1970s. By the 1980s prices and price-supply relationships in the international oil market had deteriorated, the number of buyers and sellers had increased, and market stability had turned to chaos. Increasing uncertainty about market conditions offered states opportunities, if anything, to exploit oil to their own advantage as entrepreneurs.

The "obsolescing bargain," according to Vernon, captures the repercussions of these product-cycle and market changes for bargaining between host governments and multinational companies. As long as there was a glut of oil on the market, multinational companies were in a strong bargaining position. But when shortages appeared, the balance of power shifted in favor of host governments. The obsolescing bargain identified this loss of bargaining power by multinational corporations with the erosion of their global marketing capabilities.

Vernon's analysis gives an extremely convincing view of the economic conditions necessary for state oil entrepreneurship. We can consider neither these conditions nor their bargaining consequences, however, sufficient in themselves to generate state oil enterprise. Institutional conditions also had to be conducive.

Other political theorists argue that international organizations and rules, use of force and distribution of power, constrained the ability of states to adopt roles as entrepreneurs. With their positions in the international system, Stephen Krasner argues, LDC states have two strategies to improve their position. They can win within the existing rules of the international game by gaining new resources ("relational" power); or they can change the rules of the international game by creating new organizations such as OPEC or state-owned corporations. Peter Cowhey also views OPEC as a deft new management strategy on the part of oil producers to use price setting as an oil allocation rule. In addition, he claims, members of the organization agreed mutually to adjust oil operating standards as a way to legitimize the nationalization of foreign oil holdings during the 1970s.[1]

Both of these resource and rule-changing strategies helped less developed countries legitimize more authoritative rather than market-oriented modes of allocation in the world economy. So did the UN Conference on the Law of the Seas treaty, which transferred more sovereign control and industrial wealth to LDCs than they could ever have expected through the market. But the success of LDC gamesmanship has depended upon changing the international rules of the oil game. Existing global institutional arrangements have indeed shifted from great-power primacy to sovereign equality. LDCs have learned how to put industrialized countries on the defensive in international negotiations. Finally, U.S. domination has eroded, and the vacuum been filled by European, Japanese, and Soviet power.[2] What we need to do, however, is explicitly to tie these rule-changing strategies at the international level to those of "strengthened" states at the national level.

The burden of responsibility for oil nationalizations and state entrepreneurship seems to fall on the weak international regime of force and sanctions during the 1970s and 1980s. Little was done to prevent a wave of expropriations of direct foreign investment in oil as well as other industries. As industries matured through the product cycle, and state capacity and nationalism grew, direct multinational corporate investments were expropriated. International groups failed to intervene, impose economic sanctions, or exert pressure through international banks. Only portfolio investments remained secure, as Charles Lipson argues, because host governments still needed capital; international banks could threaten severe collective sanctions if countries defaulted on loans. Such threats could also have been employed by foreign oil companies— had they not acted independently in negotiations.[3] What is notable is the generally passive role that multinational oil companies played in opposing state entrepreneurship at the national level, in stark contrast to the active role of international banks during debt crisis.

This book supplements that international viewpoint on the rules, organizations, and distributions of power governing the behavior of the oil game with a national perspective on the role of the state and domestic industry. I have focused on how state oil enterprises in both developed and less developed oil-exporting countries changed the institutional rules of national oil games. But international oil games also contribute to explaining the successes and failures of state efforts to restructure international marketing and transportation. Product-cycle and market conditions constrain both the national and the international operations of state enterprises. Because of my primary focus on the state, I have treated the roles of multinational oil companies, international banks, and home governments as international constraints within the national policy-making context.

State entrepreneurship changed the rules of the oil game at the national level. Instead of their traditional taxing of foreign oil companies, I argue, governments supplemented their tax revenues with state-earned oil profits and their aid packages or development loans with bank loans made directly to state oil companies. States also wielded substantial administrative and corporate discretion over foreign and domestic competitors in their national oil economies, simultaneously allocating and bidding for oil leases and contracts. But the role of states in national oil depended upon their ability to pursue their own autonomous interests rather than just the interests of others.

THE STATE'S CAPACITY TO ACT IN ITS OWN INTERESTS

The problem of state autonomy has intrigued statist and dependency theorists alike. Both groups ponder whether states can be analyzed as independent of societal pressures, which pits two concepts of government against each other: the centralized planner against the decentralized societal representative. The essential question for statists is whether the state can have autonomous interests independent of societal influence and can effectively implement those interests through its policies. The essential question for dependency theorists is whether the state's own interest represents a societal coalition of interests which guides national political culture and policy into a dependency on international processes of capital accumulation.

Both of these approaches focus on the state as a goal-oriented, rational actor pursuing policy change. They differ substantially from what Theda Skocpol has termed the Tocquevillian approach, which emphasizes the sociopolitical rather than the policy consequences of the state as an organization influencing group formation and issue raising within the national political culture.[4]

Mine is a variation on the policy-oriented statist approach. In brief, this book argues that the state does have its own institutional interests, which are independent of society's interests whether representative of international or of national groups. Entrepreneurial state oil policy can continuously implement the institutional interests of the state—if there is a consistency between the sovereign rights, institutional and agency interests, and political coalitions of state agencies and state enterprises. State consistency, but not necessarily state unity, is essential to autonomous action by the state bureaucracy. The argument thus has two parts: first, the state has autonomous institutional interests; second, the state can implement those interests consistently through state enterprise.

The first argument, for state autonomy, is based on evidence drawn primarily from Norway, Britain, Indonesia and Malaysia which indicates that governments used state enterprise to achieve specific bureaucratic and nationalistic goals. State oil enterprise enabled bureaucrats to extend the management control of the state as a rational planning institution, along lines proposed by Max Weber. State enterprise also increased bureaucratic budgets as well as state corporate purses by supplementing government tax revenues with direct profits, which was more effective than trying to bargain with the executive or representative branches for budget increases from the treasury. State enterprise also enabled state leaders to increase their discretionary control over territory and industry by providing a new form of sovereign authority. State companies could exercise property rights analogous to "private" ones by acquiring leases and contracting to either public or private companies, which was a more powerful way of extending state property rights than merely increasing the amount of territory that the state allocated for private property rights. Finally, state leaders could use state enterprises to serve nationalistic interests in developing oil for the nation not for foreign oil companies, a point reminiscent of dependency theory. I argue, however, that autonomy not dependency can result from partnerships with international capital.

These state interests differed from other interests associated with government. A combination of national interests in nationally owned oil development and public interests in public control of the economy and natural resources legitimized state institutional interests in oil. The state's interest, as the term is used here, was defined by how bureaucrats interpreted these national and public interests over time. What differentiated the state's interest from the bureaucracy's interest was its national industrial rather than purely civil service or administrative focus. The state's interest remained constant, moreover, despite changes of government administration.

The potential for state autonomy, I believe, is not in perpetual flux over time but instead exhibits fixed, structural characteristics. Representative pressures of governance or changes in relations to societal groups do not reduce state institutional interests to a mere vacillation. This is not to say that historical and structural variations within given polities do not influence state autonomy, of course, but it is to argue that institutional interests of the state have autonomous structural characteristics that survive those influences.

State interests, despite this structural characterization, evade unequivocal empirical verification. We cannot be certain whether officials are acting according to their own personal preferences or in accordance with their official responsibilities. Do we judge, say, Indonesian investments of the state as entrepreneur by whose institutional interests they serve, or do we accept the proposition that the state's interest is really the collective personal preference of the individuals running the government? I find the former more convincing in explaining the growth of state enterprises in oil. Short-term personal financial preferences, for instance, could be much better served by taxation or budget increases than by state enterprise, which reinvests profits and thus lowers payoffs to individuals.

The second argument is that the state can consistently pursue its autonomous institutional interests through state oil enterprise. To do so, however, the state must maintain three conditions of consistency between state interests and entrepreneurial actions.

The state must be willing to grant sovereign property rights to the state company, among them ownership of subseabed and onshore oil, the use of eminent domain to force private companies to relinquish leases, and preferential leasing rights. The extension of sovereignty to the state company integrates state rights with policy actions concerning resources and territory. Second, new corporate capacities of state enterprises must fulfill the objectives of state bureaucracies. If part of the oil profits or industrial control gained by the state enterprise does not increase the Ministry of Finance's budget or managerial discretion, for instance, the interests of ministry and enterprise will diverge. And third, an advocacy coalition must form within the state. Bureaucratic initiatives may be the basis for state autonomy in oil, but they require coalition building with executive and representative branches of the state.

Given these three conditions, I argue, some autonomous and consistent state oil interests may be implemented over time by all or some of the ministries and state enterprises in a "strong" state. But other agencies and parliaments of strong states may reflect eclectic or conflicting goals, other bureaucratic purposes (such as minimizing the public debt), or societal interests that oppose state enterprise. The more inconsistent the interests and entrepreneurial goals of a state, the less autonomous that state becomes.

Two major controversies arise over the claim that state autonomy and policy consistency are viable. The first concerns feuds within the state. The second focuses on bargaining concessions that the state must make to societal groups.

Do feuds erode the state's ability to pursue unified or consistent sets of interests? When state oil companies act contrary to interests of ministries of finance, is the state's interest still being served? States have their own institutional interests in national oil development, I find, interests that are not necessarily fragmented by the budgetary needs of individual state agencies. State agencies also have their own particular budgetary or sectoral interests, but these do not overwhelm the state's interest as long as a strong, supportive coalition retains central control within the state. Evidence is provided by overriding ministerial support for state oil enterprise in the four countries primarily analyzed in this book.

One could substitute the pluralist assumption that states have only diverse and eclectic interests represented by their ministries; ministers are often in conflict with directors of state oil companies because each has countervailing interests. Alternatively, one could make the bureaucratic assumption that agencies seek to increase their budgets by bargaining with other state bodies or by taxing the property rights that they extend to private companies. The presence of a state enterprise would provoke such bureaucratic agencies to a constant competition to bargain with executive or representative bodies for payments of profits into agency budgets. Both pluralist and bureaucratic perspectives deny the premise that the state has overriding institutional interests that can be pursued by consistent actions among state entrepreneurial and bureaucratic, executive and representative bodies. Instead, they perceive the state as constantly rent by feuds over competing interests.

But treatment of feuds within the state is not the only theoretical controversy. Statist and dependency theorists in effect reject the viability of either state autonomy or consistency, because of the bargaining concessions that the state must make to societal groups. Despite the state orientation of their arguments, even some statists have argued that the state is "neutralized" or must "resist" the opposition of societal groups. Peter Katzenstein contends, for instance, that it was precisely the interpenetration of the Austrian state bureaucracy by domestic political parties and interest groups which neutralized the state's role in nationalized industry.[5] Similarly in other small European countries, such as Norway and Switzerland, Katzenstein argues, democratic corporatist relationships bind state and society. Ideologies of social partnership among state, private business, and labor groups are strong. A centralized and concentrated core of interest groups and informal, voluntary bargaining also work to achieve consensus among interest groups, the state bureaucracy, and political parties.[6] The result is integrated and well-adjusted national political economies; the implication, however, is that state bureaucracies cannot initiate autonomous policies of their own.

Dependency theorists also find the autonomy of the state problematic. In oil and other industries in Brazil, Peter Evans argues, the state juggled three roles, defeating the success of any one of them. Relations between local and multinational capital impeded the state's role as a nationalist state defending the interests of the local bourgeoisie. Nationalism and the state's own political control slowed its determination to attract foreign investors as an entrepreneurial state selling Brazil to foreigners. And the political influence of other social groups restricted the state's effort to look out for its own interests in pursuing investments.

Furthermore, many dependency theorists doubt that consistency is possible between the state and its enterprises. Public enterprise requires a degree of political independence that results in decentralization, they argue, not merely the administrative differentiation of the state, and decentralized entrepreneurial entities can be captured by divergent societal interests. A similar argument holds that the state is fragmented by narrow sectoral interests in society, which use public oil companies to serve their own purposes. State intervention through public enterprise thus reproduces the dispersive and conflictual tendencies in society. According to Dietrich Rueschemeyer and Evans, this undermines "the state's capacity for coherent corporate action."[7]

Both the statist and the dependency models of state-society relations portray the state as so intertwined in its dealings with domestic and foreign groups that it can act neither autonomously nor consistently. In contrast, the evidence of sustained executive and bureaucratic support for state oil enterprises demonstrates that states in eleven countries *did* act autonomously in oil, and despite pressure from the representative branch or from external society. The relatively constant pursuit of oil investment in production and related operations by state oil companies indicates, furthermore, the capacity of states for consistent policy action. This consistency occurred despite changes of government leadership or concessions to opposition groups. State entrepreneurship often provoked threats by heads of government (Britain), internal state feuds (Indonesia), or defiance by domestic and foreign groups (Norway and Malaysia respectively); but those various threats did not entirely curb the state's autonomous and consistent forays in oil policy.

U.S. foreign investment policy supports this argument for state autonomy but only on a sectoral basis. Krasner argues that the bureaucratic insulation and obligation to serve national interests of the president and secretary of state provided the basis for a certain degree of autonomy in U.S. foreign policy concerning raw materials. This autonomy, however, was policy-specific within a state otherwise extremely weak because of congressional influence and interest-group fragmentation.[8]

The unevenness of state power does indeed seem to make state autonomy virtually impossible on anything but a sectoral basis. Differences in the state's role in oil, sugar, and manufacturing suggest that broader discussions of state strength and policy independence must be disaggregated into sector-specific considerations.[9] The variation in state capital resources, policy goals, technology and monitoring capabilities, and international position is likely to be too great across sectors for unified analyses of state capacity. In fact, sectorally based oligopolistic groups led by state enterprises may segment national economies into semi-autonomous sectors, united only at the highest level of policy making.[10]

The variation in state autonomy across sectors is an important point. In this book, however, I have avoided overinterpreting the role of oil by treating opposition from other domestic industries, shipping and fishing, as an intervening variable. As a result I portray state policy outcomes in oil as reflecting financial, operational, and territorial policy concessions to other industrial sectors. In this way I integrate policy outcomes across sectors in relation to the overall autonomy of the state within a single sector, oil.

BARGAINING CONSTRAINTS ON THE EXPANSION OF STATE ENTERPRISE

Bargaining has been popular as a way to understand the give and take of governments responding to pressures from outside groups. International groups include multinational corporations, international banks, and foreign governments, while domestic groups consist of private national companies, political parties, labor unions, industry associations, and other interest groups. As bargaining models have evolved, they have presented windows on the changing forces that constrain state enterprise.

Early state-centric views suggested that host governments offered multinational companies access to territory and resources in return for rent in the form of tax revenues. This notion was later extended to include a role for foreign companies in national contracts for the provision of goods and services. State-multinational bargaining also helps explain why the multinationals could not prevent expropriations of their direct investments. Once states had expanded their own production capacities and had been spurred on by nationalistic fervor, they could be stopped only by concerted international sanctions or force.

But the conceptual limits of state bargaining with multinationals were particularly apparent in Latin America, where domestic groups also had interests in nationalizing foreign investments in raw materials. Theodore Moran rejected the state-centric model, arguing that the Chilean state's move to nationalize copper resulted from the pressure of domestic groups to stem a growing dependency of the national economy on foreign companies. Copper policy in Chile, according to Moran, was "the outcome of the interplay of domestic groups trying to maximize their own particular interests as well as the larger national interest."[11] He thus substituted domestic groups for the state as the key local determinant of national policy. Moran explained the eventual nationalization of the major foreign mining companies—Kennecott, Cerro, and Anaconda—by Salvador Allende in 1971 by focusing on bargaining between multinational corporations and domestic groups over concession contracts.

Moran then linked product-cycle to bargaining theory by arguing that investment uncertainty and the learning curve of domestic groups determined bargaining outcomes. Once investments were sunk, and declared successful, bargaining turned to contract re-negotiations that favored domestic groups instead of the monopoly control of multinational corporations. Similarly, domestic groups gained negotiating and industrial expertise as the industry ma-tured through the product cycle. Both conditions favored eventual nationalizations led by Chilean groups.

But other dependency and statist theorists found neither state-multinational nor domestic-multinational bargaining models suffi-cient to explain industrial policy outcomes in Latin America and Europe. They needed to break "society" analytically into domestic and international influences, which resulted in interpretations of three-way bargaining between state, multinational, and domestic groups to explain policy outcomes. But did this bargaining repre-sent a growing dependence of the state and domestic groups on international capital or a consensus building that passified state initiative?

Within the "dependent development" of Brazil, Evans argued, was a triple alliance between elite local, international, and state capitalists, all of them working to accumulate foreign capital within the country. Despite their potentially diverging interests, the mem-bers of this alliance had much to gain collectively from drawing Brazil into a path of dependent development reliant on interna-tional capital. Fernando Cardoso and Enzo Faletto relied on the same bargaining dependency model but emphasized that domestic groups played different roles depending upon the origins of the state in their countries. The populist roots of the state in Brazil and Argentina made it hard for governments to control the national bourgeoisie and local companies. As a result, military coups en-abled techno-bureaucrats to repel democratic power and replace democratic-representative regimes with authoritarian-corporative regimes. Mexico, by contrast, started industrialization with the state already an investor and regulator, resulting in a new kind of oligarchy (of state, multinational, and local capital) that reflected Evans's triple alliance.[12]

The drawback of the dependency bargaining model is that it portrays state actions as reflecting the dependence of the state and other groups on furthering the interests of international capital. This is true for state enterprise as well as for national industrial policy more generally. This bargaining model seeks to explain not the degree of development but the degree of dependency in development.

Statist theorists have avoided predetermining the bargaining outcome, not by allowing for autonomous state policy choices but by showing how consensual three-way bargaining diverts outcomes away from the path chosen by the state. Katzenstein argues that cooperative political arrangements in the small countries of Western Europe rely on consensual bargaining among the state, private business, and labor groups. International business and diplomatic forces provoke a collective effort by state and domestic groups to withstand competitive pressures and the shocks of the international economy. International business interests are incorporated into domestic business concerns, and the consensual three-way bargaining that results is based on an informal, voluntary commitment to use bargaining as the means to achieve cooperation.

In Austria such democratic corporatist bargaining weakened a relatively strong, centralized state. Strong political parties reduced the state to a passive entrepreneur in nationalized industry and a compliant leader of nationalized banks. But the same kind of consensus politics led to an opposite outcome in Switzerland. The Swiss state, though weak and decentralized, needed to arbitrate and unify the bureaucracy, unions, and internationally oriented business; the result was a strengthening of the state.[13] The increasing need of small countries to offset political and economic disruptions caused by their openness to the international economy has fostered this strong interpenetration of state, domestic, and multinational groups in policy making.[14]

Both dependency and statist bargaining analyses employ three-way bargaining models to explain consensus building among state, multinationals and domestic groups. Neither approach presents state autonomy as a viable outcome of bargaining. The two approaches differ in that dependency theory sees international capital propelling the triple alliance into dependency, while statist theory sees state-multinational-domestic consensus building as a variable but healthy response of national economies to shocks from the international economy. In fact, Katzenstein regrets that large countries—the United States, Britain, Japan, France—do not exhibit the sturdiness of democratic corporatist bargaining. In such large countries the state bureaucracy and political parties remain decentralized; strong liberal ideologies and political forces continue to favor market rather than coalition-building solutions to international disruptions.

I also use a comparative, three-way bargaining analysis, but I explicitly include bargaining within the state itself. I assume state autonomy and then examine the extent to which bargaining constrains that autonomy in state oil enterprise. Bargaining first occurs among executive, bureaucratic, representative, and entrepreneurial branches of the state itself and then among state, domestic, and multinational groups. Hence the state can pursue autonomous interests despite periodic threats from internal or external state opponents.

The internal state bargaining among bureaucratic, executive, representative, and entrepreneurial bodies is thus not the only support for or constraint on autonomous state oil policies. External domestic and international groups can and do threaten the state coalition. State coalitions backing state enterprise will be threatened to the extent that domestic groups—local industrialists, labor, environmentalists, fishermen, and others—erode support in representative branches, for instance parliaments. Such was the case in the advanced industrialized countries, which also suffered from threats or the desire to avoid threats imposed by multinational groups (oil companies, international banks, and foreign governments). Domestic groups were so politically weak in less developed countries that the LDC states had to deflect threats only from these multinational groups. Needing to bargain relatively less to offset opposition from external groups, LDC states were more effective than advanced industrialized states in pursuing autonomous, entrepreneurial oil policies.

The notion of an autonomous state, periodically constrained by bargaining threats from domestic and international groups, is the basis for a comparative logic of statist bargaining in oil-industrializing countries. State oil entrepreneurship in advanced industrialized countries such as the United States, Britain, France, and

Japan was constrained both by strong domestic coalitions and by strong international coalitions. This domestic-multinational opposition is the basis for the combined domestic and multinational concession pattern identified in most advanced industrial countries. The state-multinational-domestic bargains found in Brazil and Argentina broaden my argument by suggesting that the presence of strong domestic coalitions, whether in advanced industrialized or less developed countries, leads to a weaker entrepreneurial role for the state.

In contrast, state oil entrepreneurship in less developed countries such as Indonesia, Malaysia, Mexico, Iran, and Saudi Arabia flourished. LDC states were constrained, primarily by international coalitions, but mostly at the international level of operations. Domestic coalitions were too weak to halt state-run operations. Concessions to multinational corporations, international banks, and foreign governments established only a multinational concession pattern in most LDCs. Norway is an unusual case in that it exhibits primarily a domestic concession pattern, while Italy exhibited the multinational concession pattern found in less developed countries even though it is an advanced industrialized country.

This three-way bargaining perspective differs from other approaches by assuming that opposing groups only sometimes jeopardize actions of autonomous states. Statist and dependency approaches, by contrast, view states as being so intertwined with societal groups that their interests and actions are almost always compromised. My approach also differs from other oil industry analyses in offering a systematic analysis of comparative state growth and domestic-multinational opposition in eleven countries.

The comparison of domestic and multinational opposition to the state also provides more of a perspective on bargaining internal to the nation state than do product-cycle and other international bargaining approaches. I combine a bureaucratic and statist approach to international bargaining, focusing primarily on the new political resources of the state and its opponents at the national level. Other bargaining approaches emphasize existing market conditions, property rights, or rules that facilitate or constrain the growth of state enterprises at the international level.

One potential point of contention is my treatment of state autonomy. By assuming that the state is analytically separate, and *not* inextricably bound to society, I recognize the bargaining autonomy of state leaders. To assume state-society interdependence, as do many statist and dependency theorists, would be to assume that the state could exercise only constrained bargaining freedom, because we would have to assume that societal influences are always present. Such interdependent models would need a truly independent measure of autonomy if the state were ever to have its own way entirely in a win-loss bargaining outcome.

Another potential concern is my implicit rather than explicit theoretical treatment of partisan and elite politics. I neglect partisan politics as an independent variable even though political parties and ideologies have clearly been important factors in both Norway and Britain. The reason is that the comparative structure argument I wish to make relies on the interplay of state, domestic, and international actors to explain policy outcomes. I have tried, therefore, to avoid explaining oil policies as the results of particular ideologies or strategies, of left- or right-wing parties in advanced industrialized countries and technocratic-military elites in less developed countries. Hugh Heclo's work on social policy in Britain and Sweden supports this stance in arguing that civil servants had much greater influence on policy than did political parties, interest groups, or state legislatures. In particular, Sweden's centralized bureaucratic state was more effective than the British state, even after the democratization of national politics, in autonomous policy formation.[15] Centralized administrative bureaucracies with autonomous agendas in Germany, as in Sweden, preempted the vote getting and patronage wielding of political parties which grew in the United States to fill the political vacuum of a decentralized, federal bureaucracy.[16]

But political parties are important in small countries such as Norway and Austria. Katzenstein argues that partisan politics there play the crucial role of narrowing differences between interest groups. In large countries such as Britain, on the other hand, this role of political parties was less important. Political parties remained decentralized, and liberal ideologies continued to favor market solutions rather than the building of political consensus.

Other work fills out the political story in such LDCs as Indonesia and Peru. The technocratic-military elites that govern Indonesia play a strongly centralized role, and political parties and parliaments play roles subordinate to rule by those elites. The socialization of new military professionals in Peru was a precondition for the corporatist coups that took place in the 1960s. These career military officials were trained in national economic planning and counterinsurgency, and so they could incorporate state planning and socioeconomic reforms into their efforts to deflect political threats to national order.[17]

CONCLUSION

In this book I have provided a comparative national perspective on oil politics, particularly from 1960 to 1985. The book complements product-cycle and international bargaining theory that focuses mainly on international or transnational relations among multinational corporations, international banks, and host and home governments. By examining the role of the state mainly at the national level, I bring into focus the organizational struggles of international actors within the arena of domestic business, labor, political parties and interest groups. I thus supplement other analyses of domestic politics by statists as well as by dependency theorists. But in focusing on the bureaucratic and entrepreneurial aspects of the state in oil, I have also drawn heavily from bureaucratic theory on the politics of management, budget, and the allocation of property rights within the state itself.

This institutional focus is the basis for a new perspective that emphasizes the capacity of state bureaucracies to pursue autonomous and consistent policy objectives through state enterprise. My research across eleven countries demonstrates this capacity for state autonomy in oil and suggests that state enterprise can be an effective industrial planning tool for capitalist governments. The research also shows that domestic groups constrained states using state enterprise more in advanced industrialized countries than in less developed countries. Both sets of countries, however, were constrained by international groups exercising their control to reinforce product-cycle barriers to market entry.

State enterprise, it appears, can be more effective as an industrial policy for centralized capitalist governments in the developing world than it can ever be for most decentralized, representative governments in industrialized Western countries. The various coalition-building politics of Western countries, whether modified market or democratic corporatist, favor policy packages that parcel out industrial operations to contending state, domestic, and multinational groups rather than reserving them for the state alone.

Notes

CHAPTER 1. *The Struggle for Control of National Oil*

1. See Hansan Zakariya, "State Petroleum Enterprises: Some Aspects of Their Rationale, Legal Structure, Management, and Jurisdiction," in United Nations Centre for Natural Resources, Energy, and Transport, *State Petroleum Enterprises in Developing Countries* (New York: Pergamon, 1980), p. 32.

2. Seventy-eight countries in the world had partly or wholly state-owned oil companies by 1984. U.S. Department of Commerce, International Trade Administration, *Government-Owned Oil Companies around the World*, ed. Charles M. Cummings, Sr. (Washington, D.C., 1984).

3. Stale Seierstad, "The Norwegian Economy," in Natalie Rogoff Ramsøy, ed., *Norwegian Society* (New York: Humanities, 1974), pp. 91–92, and Fritz Hodne, *The Norwegian Economy, 1920–1980* (New York: St. Martin's, 1983).

4. Seierstad, "The Norwegian Economy," p. 92.

5. Norwegian Shipowners Association, *Review of Norwegian Shipping, 1977* (Oslo, 1977).

6. Seierstad, "The Norwegian Economy," p. 77.

7. Øystein Noreng, *Petroleum and Economic Development: The Cases of Mexico and Norway* (Lexington: Heath, 1984), p. 5.

8. Economist Intelligence Unit, *Quarterly Economic Review [QER] of the United Kingdom*, Annual Review, 1967, pp. 1–2.

9. *QER of Indonesia*, Annual Supplement, 1968, p. 2.

10. Ibid., Annual Supplement, 1972, p. 2.

11. Ibid., 1974, no. 1, p. 2; 1975, no. 3, p. 2.

12. Ibid., 1975, no. 2, p. 3.

13. Ibid., 1976, no. 1, pp. 2–3.

14. Ibid., 1982, no. 2, p. 5.

15. Ibid., *Annual Supplement*, 1983, p. 4.

16. Yahya A. Muhaimin, "Indonesian Economic Policy, 1950–80: The Politics of Client Businessmen" (Ph.D. Diss., MIT, 1980), p. 59.

17. Ibid., p. 60, and *QER of Indonesia*, Annual Supplement, 1983, p. 5.

18. *QER of Indonesia*, Annual Supplement, 1983, p. 5.

19. Ibid., 1984, p. 5.

20. U.S. Dept. of State, *Background Notes, Malaysia* (Washington, D.C., 1984), pp. 4, 5, 7.

21. See Joseph Stern, "Malaysia: Growth and Structural Change," Conference on the Malaysian Economy, Fletcher School of Law and Diplomacy, Tufts University, Medford, Mass., November 1984, pp. 3–5.

22. Ibid., pp. 27, 33.

CHAPTER 2. *A Statist Perspective on Public Enterprise in Petroleum Resource Management*

1. See Raymond Vernon, *Two Hungry Giants: The United States and Japan in the Quest for Oil and Ores* (Cambridge: Harvard University Press, 1983), pp. 26–34.

2. James Griffin and David Teece, *OPEC Behavior and World Oil Prices* (London: Allen & Unwin, 1982), p. 5.

3. Thomas Neff, "The Changing World Oil Market," in David Deese and Joseph Nye, eds., *Energy and Security* (Cambridge, Mass.: Ballinger, 1981), pp. 26–27.

4. Cf. Robert O. Keohane and Joseph S. Nye, Jr., *Power and Interdependence: World Politics in Transition* (Boston: Little, Brown, 1977).

5. Richard K. Ashley, "The Poverty of Neorealism," *International Organization* 38 (Spring 1984), pp. 225–85.

6. Robert Gilpin, *War and Change in World Politics* (Cambridge: Cambridge University Press, 1981), p. 19.

7. Malcolm Gillis, "The Role of State Enterprises in Economic Development," *Social Research* 47 (Summer 1980), pp. 252–53.

8. Max Weber, *The Theory of Social and Economic Organization* (New York: Oxford University Press, 1946), pp. 56–58, 156.

9. The notion of autonomous state interests used here differs somewhat from the definition proposed by Eric Nordlinger. He defines autonomy as the ability of any social entity to translate policy preferences into actions. A state is autonomous to "the extent that public policy conforms to the parallelogram . . . of the public officials' resource-weighted preferences." These preferences are based on the individual as the unit of analysis and represent unified "state preferences" only when a single body of the state—such as a city council—achieves an unusual consensus among countervailing career interests or organizational loyalties, or within the professional knowledge of public officials. Nordlinger, *On the Autonomy of the Democratic State* (Cambridge: Harvard University Press, 1981), p. 8.

10. Weber, *Theory of Social and Economic Organization,* pp. 337–38.

11. Øystein Noreng, *The Oil Industry and Government Strategy in the North Sea* (London: Croom Helm, 1980), p. 25, and Merrie Klapp, "Policy and Politics of North Sea Oil and Gas Development," in Jonathan Aronson and Peter Cowhey, eds., *Profit and the Pursuit of Energy: Markets and Regulation* (Boulder: Westview, 1983), p. 112.

12. Alan Budd, *The Politics of Economic Planning* (Manchester: Manchester University Press, 1978), p. 122.

13. Klapp, "Policy and Politics," p. 119.

14. Merrie Klapp, "Industrial Policy Offshore: The International Boundaries of State Enterprises," *Journal of Comparative and Commonwealth Politics* 22 (March 1984), pp. 33–38.

15. See Friedrich Kratochwil, "On the Notion of 'Interest' in International Relations," *International Organization* 36 (Winter 1982), p. 26.

16. Bruce Andrews, "Response to Ashley," *International Organization* 38 (Spring 1984), p. 323.

17. Peter Katzenstein, "Capitalism in One Country? Switzerland in the International Economy," *International Organization* 34 (Autumn 1980).

18. Peter Katzenstein, *Small States in World Markets* (Ithaca: Cornell University Press, 1985), pp. 24–58.

19. Peter Katzenstein, "Small Nations in an Open International Economy. The Converging Balance of State and Society in Switzerland and Austria," in Peter Evans, Dietrich Rueschemeyer, and Theda Skocpol, eds., *Bringing the State Back In* (Cambridge: Cambridge University Press, 1985), pp. 231–45; Katzenstein, *Corporatism and Change* (Ithaca: Cornell University Press, 1984).

20. Stephen D. Krasner, *Defending the National Interest: Raw Materials Investments and U.S. Foreign Policy* (Princeton: Princeton University Press, 1978), p. 11.

21. Stephen Krasner, *Structural Conflict* (Berkeley: University of California Press, 1985), pp. 3–15.

22. John Zysman, *Governments, Markets and Growth* (Ithaca: Cornell University Press, 1983), pp. 75–76, 86.

23. Stuart Holland, *The State as Entrepreneur* (London: Weidenfeld & Nicolson, 1972), pp. 3–20.

24. John Freeman, "State Entrepreneurship and Dependent Development," *American Journal of Political Science* (February 1982), pp. 92–93.

25. Effective state action could be measured by the extent that the state provides for the welfare and protection of society—not itself as an institution (Gilpin, *War and Change in World Politics,* pp. 16–17). But this normative, public-goods notion of public management forecloses the empirical possibility that the state can manage the national economy independent of its ability to redistribute gains to societal groups. Governments in Norway, Britain, Indonesia, and Malaysia actually made oil investments that first served their managerial interests; only secondarily did they con-

cede gains to domestic groups as political payoffs. These payoffs were *not* the government's primary intention in making oil investments.

26. Raymond Vernon, "Linking Managers with Ministers: Dilemmas of the State-Owned Enterprise," *Journal of Policy Analysis and Management* 4, 1 (1984), pp. 39–55.

27. Ibid., pp. 42, 50.

28. The original French definition of sovereignty, *souveraineté* was supreme power, deriving from the Latin *superanus*. The French Constitution of 1791 stated that "Sovereignty is one, indivisible, unalienable and imprescriptible; it belongs to the nation; no group can attribute sovereignty to itself nor can an individual arrogate it to himself." This attribution of supreme power to the nation-state was modified by the British state. Sovereignty was instead vested in the crown, but Parliament could enact laws binding everyone else but itself. Still, this legislative sovereignty sustained the notion of supreme power in the name of the state. Finally the Constitution of the United States placed sovereignty in itself as a document, retaining supreme authority but giving the people the right to approve changes in that authority according to the Tenth Amendment.

29. Heinrich von Treitschke, excerpt from *Politics* (1916), published in Richard Cox, ed., *The State in International Relations* (San Francisco: Chandler, 1965), p. 54.

30. Douglas North, *Structure and Change in Economic History* (New York: Norton, 1981), pp. 17–38.

31. Norwegian Ministry of Industry, *Report No. 30 to the Norwegian Parliament* (1973–74) (unofficial government translation), p. 35.

32. The British Petroleum and Submarine Pipelines Act received royal assent in November 1975.

33. Anderson G. Bartlett, et al., *Pertamina: Indonesian National Oil* (Djakarta: Amerasian, 1972), p. 181.

34. Malaysian Ministry of Finance, *Economic Report 1977–78* (Kuala Lumpur, 1978), 6:114.

35. William Niskanen, *Bureaucracy and Representative Government* (New York: Aldine-Atherton, 1971), pp. 5–10, 20–40.

36. Douglas Bennett and Kenneth Sharpe, "The State as Banker and Entrepreneur," *Comparative Politics* (January 1980), p. 183.

37. Bartlett et al., *Pertamina*, pp. 297–99.

38. This notion of "new" state interests does not suggest that any act of the state reflects the state's interest in expanding its bureaucratic authority or corporate powers. Coalitions of ministers and bureaucrats which control the state during any particular government administration interpret the state's interest through their choices of oil policies. If one minister or bureaucrat has overriding power over oil policy choices, then the issue is whether that official's interpretation of the state interest represents his personal preference or his bureaucratic authority. Such was the case in Indonesia, particularly during 1968–74. Ibnu Sutowo, head of Pertamina

and minister of mines, held almost exclusive control over the state oil company's investments, the reinvestment of foreign oil company tax revenues, and the share of oil tax revenues received by the Indonesian Ministry of Finance. His own interpretation of the state's interest in oil was thus translated directly into state action. This, a reviewer of this book comments, is the "private appropriation of public goods as well as the undermining of the state's form of domination," and thus an undesirable but not inconsistent pursuit of the state's interests.

One man who dominates state oil policy could also, according to Weber's notion, reflect the power of an official to act on behalf of the state, using its authority. This could be interpreted as a consistent pursuit of state interests. If these actions are judged to be corrupt, as the foreign press did in 1974–75 when Pertamina's $10 billion foreign debt was revealed, then the issue of good and bad interpretations of the state's interest is critical to that judgment. Does a corrupt or unwise interpretation of the state's interest by an official—such as accruing a large foreign debt—no longer represent the state's interest or its legitimate authority of public action? In a democratic society the answer is found in the process of public representation. When and if the electorate, representative bodies of the state such as parliaments, or judicial bodies decide that the state's interest has been "misinterpreted" or mismanaged by any high-ranking public official, they will remove him or her from office. In less representative societies this responsibility is left to the president or prime minister and his or her ministers. This was ultimately the case in Indonesia. The burden of responsibility to judge the correct interpretation of the state's interest in state enterprise, therefore, ultimately vests either in the electorate and representative bodies or in the executive of the state.

39. Merrie Klapp, "Inter-Industry Conflict in the North Sea and South China Sea: A Comparative Analysis of Oil, Shipping, and Fishing in Four Nations" (Ph.D. diss., University of California, Berkeley, 1980), p. 101.

40. See Samuel Huntington, "Transnational Organizations in World Politics," *World Politics* 25 (October 1972–July 1973), pp. 333–68, and Barbara Haskel, "Access to Society: A Neglected Dimension of Power," *International Organization* 34 (Winter 1980).

41. *Norwegian–United Kingdom Agreement Relating to the Delimitation of the North Sea between the Two Countries*, London, 10 March 1965 (entered into force 29 June 1965), UKTS no. 71 of 1965 (Cmnd. 2757). The 1969 agreement between Indonesia and Malaysia delimited national boundaries of the continental shelf covering the Straits of Malacca and the South China Sea. Similar agreements were signed between these two countries and Singapore and Thailand between 1971 and 1973. See *Convention on the Continental Shelf*, Geneva, 29 April 1958; entered into force 10 June 1964.

42. Haskel, "Access to Society," pp. 111–17.

43. Klapp, "Inter-Industry Conflict," chaps. 2–5.

44. Keohane and Nye, *Power and Interdependence*, p. 55. If exploration

or production requirements of states were too stringent, multinational corporations could declare fields uncommercial or leave to produce oil in countries offering better terms on licenses. State participation was therefore minimal. Lenient profit-sharing or carried-interest agreements were concluded in Britain and the more financially secure Norway. In Indonesia the government had almost no participation through work contracts. This state dependence upon maintaining national production levels left governments with relatively small shares of returns from national resources. By the early 1970s the Norwegian government was receiving only 20 percent of foreign company earnings through taxes and royalties. The Indonesian government could only claim that 30 percent of its income was from oil revenues. Thus the bargains reached between governments and foreign multinational oil companies during the 1960s and early 1970s were based on asymmetrical needs. Governments needed the capital, expertise, and technological capacity of foreign multinational oil companies to develop national oil industries. In return they could offer access to state-controlled territory and resources, which foreign companies could also get elsewhere. See Norwegian Central Bureau of Statistics, *The Oil Activities on the Norwegian Continental Shelf up to 1977* (Oslo, 1978), p. 15, and *Industrial Statistics, 1976* (Oslo, 1978); Bruce Glassburner, "Indonesia's New Economic Policy and Its Sociopolitical Implications," in Karl Jackson and Lucien Pye, eds., *Political Power and Communications in Indonesia* (Berkeley: University of California Press, 1978), p. 153.

45. Klapp, "Inter-Industry Conflict," chaps. 2–5; Noreng, *The Oil Industry*, p. 150; and Lind and MacKay, *Norwegian Oil Policies*, p. 62.

46. Theodore Moran, "Multinational Corporations and Dependency: A Dialogue for Dependentistas and Non-Dependentistas," *International Organization* 32 (Winter 1978), p. 82; C. Fred Bergsten, Thomas Horst, and Moran, *American Multinationals and American Interests* (Washington, D.C.: Brookings, 1978); and Bergsten, "Response to the Third World," *Foreign Policy* no. 17 (Winter 1974), pp. 3–34.

47. Fariborz Ghadar, *The Evolution of OPEC Strategy* (Lexington: Heath, 1977), p. 35.

48. Raymond Vernon, "*Sovereignty at Bay* Ten Years After," *International Organization* 35 (Summer 1981), p. 527.

49. Cf. Vernon, *Two Hungry Giants*, pp. 35–37.

50. Katherine Huger, "North Sea Oil Development Policy: A Case Study of the Government-Industry Relationship in Norway and the United Kingdom," *Fletcher Forum* 1 (Autumn 1976), pp. 32–61.

51. Interview #147 with former high-level official in Pertamina and OPEC, Jakarta, Indonesia, 19 September 1978; interview #143 with former high-level official, Ministry of Mines, Jakarta, Indonesia, September 12, 1978. See List of Informants following Notes.

52. Huger, "North Sea Oil," pp. 32–61; interview #35 with high-level official, Economics and Legal Division, Petroleum Directorate, Stavanger,

Norway, 7 February 1978; and interview #39 with high-level official, Ministry of Commerce and Shipping, Oslo, Norway, 14 February 1978.

53. Interview #147; interview #143; interview #139 with consultant, Ministry of Finance, Jakarta, Indonesia, 8 September 1978.

CHAPTER 3. *Oil Policy Options*

1. Roderick O'Brien, *South China Sea Oil: Two Problems of Ownership and Development,* Institute of Southeast Asian Studies, Occasional Paper no. 47 (Singapore, August 1977), p. 53.

2. Merrie Klapp, "The State—Landlord or Entrepreneur?" *International Organization* 36 (Summer 1982), p. 577.

3. Norwegian Ministry of Industry, *Report No. 30 to the Norwegian Parliament* (1973–74) (unofficial government translation), and Øystein Noreng, *The Oil Industry and Government Strategy in the North Sea* (London: Croom Helm, 1980), pp. 122–23.

4. Norwegian Ministry of Industry, *Report No. 30,* p. 45.

5. Charles Johnson, "Establishing an Effective Production-Sharing Type Regime for Petroleum," *Resources Policy,* June 1981, p. 133.

6. Interview #147 with former high-level official in Pertamina and OPEC, Jakarta, Indonesia, 19 September 1978.

7. Johnson, "Establishing an Effective Regime," pp. 133–34, and O'Brien, *South China Sea Oil,* pp. 64–65.

8. O'Brien, *South China Sea Oil,* p. 65.

9. *Norwegian-United Kingdom Agreement Relating to the Delimination of the North Sea between the Two Countries,* London, 10 March 1965 (entered into force 29 June 1965), UKTS No. 71 of 1965 (Cmnd. 2757), and *Convention on the Continental Shelf,* Geneva, 29 April 1958; entered into force 10 June 1964.

10. Norwegian Ministry of Industry, *Report No. 76 to the Norwegian Storting (1970–71): Exploration for and Exploitation of Submarine Natural Resources on the Norwegian Continental Shelf* (Oslo, 30 April 1971), p. 32, and Norwegian Central Bureau of Statistics, *Industrial Statistics, 1976* (Oslo, 1978), p. 21.

11. This sketch is based on Noreng, *The Oil Industry.*

12. Norwegian Ministry of Industry, *Report No. 30,* pp. 43–44; Norwegian Central Bureau of Statistics, *Industrial Statistics, 1976,* p. 35; and Noreng, *The Oil Industry,* pp. 122, 165.

13. Norwegian Central Bureau of Statistics, *The Oil Activities on the Norwegian Continental Shelf up to 1977* (Oslo, 1978), p. 15; Norwegian Central Bureau of Statistics, *Industrial Statistics, 1976,* pp. 15, 33; and Norwegian Ministry of Industry, *Report No. 76,* p. 32.

14. Norwegian Ministry of Industry, *Report No. 30,* pp. 43–44.

15. Interview #29, 2 February 1978, with official, Public Affairs and Information Department, Statoil, Stavanger, Norway (see List of Infor-

mants following Notes); Statoil, *Annual Report and Accounts 1976* (Oslo, 1976), p. 14; and Norwegian Central Bureau of Statistics, *Industrial Statistics, 1976*, p. 19.

16. Norwegian Ministry of Industry, *Report No. 30*, p. 35.

17. M. M. Sipthorp, ed., *The North Sea: Challenge and Opportunity*, David Davies Memorial Institute of International Studies (London: Europa, 1975), pp. 262–63.

18. Interview #35 with high-level official, Economics and Legal Division, Petroleum Directorate, Stavanger, Norway, 7 February 1978, p. 50, and Norwegian Ministry of Industry, *Report No. 30*, p. 50.

19. Norwegian Shipowners Association, *Review of Norwegian Shipping, 1977* (Oslo, 1977); *Norwegian Shipping News*, nos. 9–16 (Oslo, 1971); interview #35 with high-level official, Economics and Legal Division, Petroleum Directorate, Stavanger, Norway, 7 February 1978; and interview #39 with high-level official, Ministry of Commerce and Shipping, Oslo, Norway, 14 February 1978.

20. Norwegian Ministry of Industry, *Report No. 30*, p. 78.

21. Katherine Huger, "North Sea Oil Development Policy: A Case Study of the Government-Industry Relationship in Norway and the United Kingdom," *Fletcher Forum* 1 (Autumn 1976), pp. 49, 135.

22. Interview #35, and interview #39.

23. Interview #47 with high-level official, Petroleum Legislation Division, Petroleum Directorate, Oslo, Norway, 1978, and interview #41 with executive, R. S. Platou A/S (shipping brokerage firm), Oslo, Norway, 1978.

24. Interview #35, and interview #39.

25. Statoil, *Annual Report and Accounts, 1976;* Norwegian Central Bureau of Statistics, *Industrial Statistics, 1976*, pp. 19, 31, 35; *World Oil*, 15 August 1980, p. 453. *Quarterly Energy Review, Western Europe*, 2d quarter 1981, p. 73; ibid., Annual Supplement, 1982, p. 65; Economist Intelligence Unit, *Country Report: Norway*, 1986, no. 2. The Norwegian government's share of oil and gas production value in 1980, however, was substantial: 25 billion kroner in taxes out of a total of 43 billion kroner. See Gunnar Gjerde, "Norwegian Petroleum Policy: Factors of Importance When Deciding the Extraction Rate," *Cooperation and Conflict* 17, 2 (1982), p. 96.

26. *World Business Weekly*, 13 October 1980, pp. 15–16, and 5 May 1980, p. 14.

27. *Petroleum Intelligence Weekly*, 5 January 1981, p. 7. By 1982 Norway was gaining significant market power by becoming both the only gas-exporting country and net exporter of oil in Western Europe. See Martin Saeter, "Natural Gas: New Dimensions of Norwegian Foreign Policy," *Cooperation and Conflict* 17, 2 (1982), p. 141.

28. *Petroleum Economist*, January 1980, p. 4, and August 1980, p. 4.

29. Sibthorp, *North Sea*, p. 11; Directorate of Fisheries, "Fiskerirettlederen i Haugesund-Bokn-Tyvaer-Utsina," internal document, Bergen,

1978; interview #19 with high-level official, Legal Division, Directorate of Fisheries, Bergen, Norway, 25 January 1978; and interview #25 with high-level officer, Sør Norges Tralerlag (trawlers' association for the North Sea), Karmøy, Norway, 30 January 1978.

30. Actually 9,133,563 kroner, or $1,665,049. This amount covered slightly more than 1,145 claims (out of 1,645 total) by fishermen for compensation, according to Directorate of Fisheries, internal documents.

31. Interview #24 with high-level official, Rogaland District Fishing Organization, Karmøy, Norway, 30 January 1978, and "North of 62— Ready for 1978 Action?" *Noroil* no. 1 (January 1978).

32. *World Business Weekly*, 5 May 1981, p. 14.

33. OECD, *Review of Fisheries in OECD Member Countries, 1976* (Paris, 1977), p. 186.

34. However, in 1980 drilling had begun on two fields north of the 62d parallel. By 1981 three additional blocks had been licensed in the north. *Petroleum Intelligence Weekly*, 6 April 1981, p. 10.

35. *Economist*, 18 April 1981, pp. 70–71; European Free Trade Association, *Annual Supplement* 11/79, pp. 12–13.

36. *Economist*, 18 April 1981, pp. 70–71.

37. European Free Trade Association, *Annual Supplement*, 11/79, p. 13.

38. National Westminister Bank, *Quarterly Report*, pp. 13–14.

39. *Petroleum Economist*, August 1980, p. 312.

40. Ibid., November 1980, p. 476, and January 1984, pp. 22–24.

41. Ibid., January 1984, pp. 22–24, and May 1980, p. 214.

42. Ibid., January 1984, p. 24.

43. Keith Chapman, *North Sea Oil and Gas: A Geographical Perspective* (Newton Abbott: David & Charles, 1976).

44. By 1971, the first year of production, the government had received a total of 1.37 million Norwegian kroner ($89 million) in initial payments from oil companies. See ibid., p. 14; D. I. MacKay and G. A. MacKay, *The Political Economy of North Sea Oil* (London: Robertson, 1975), p. 25; and Sibthorp, *North Sea*, pp. 255–58.

45. Noreng, *The Oil Industry*, pp. 113–20, 161–70.

46. Alan Budd, *The Politics of Economic Planning* (Manchester: Manchester University Press, 1978).

47. Evan Brown, "Finance for the North Sea: A Matter of European Concern," Royal Institute of International Affairs, Oslo, February 1975.

48. Noreng, *The Oil Industry*, pp. 121, 137.

49. BNOC was involved in thirteen of the seventeen oil fields under production or development in the British sector by 1978. Among the international oil companies involved in these production agreements were Gulf, Continental, BP, Occidental, Tenneco, Shell, and Exxon. U.K. Department of Energy, *Development of the Oil and Gas Resources of the United Kingdom, 1978*, Report to Parliament by the Secretary of State for Energy (London: HMSO, 1978), p. 21.

50. U.K. production in thousand b/d was 1973, 8; 1974, 9; 1975, 20; 1976, 245; 1977, 770; 1978, 1080; 1979, 1570; and 1980, 1619. U.S. Department of Energy. Energy Information Administration, *Monthly Energy Review,* April 1981, p. 93. The 1977 price of British crude oil is not available, but the 1978 average f.o.b. price (including insurance and transportation costs) was $13.62 per barrel. See ibid., January 1981, p. 79; U.K. Department of Energy, *Development of Oil and Gas Resources,* p. 20; MacKay and MacKay, *Political Economy,* p. 28.

51. U.K. Ministry of Trade, Board of Trade, *Report* presented to Parliament in May 1970, by the Committee of Inquiry into Shipping (London: HMSO, 1970), p. 429; Ignacy Chrzanowski, *Concentration and Centralization of Capital in Shipping,* ed. S. J. Wiater (Lexington, Mass.: Lexington, 1975), p. 30; and Central Office of Information, *Shipping* (London: HMSO, 1974), p. 4.

52. *Economist,* 9 May 1971, and U.K. Ministry of Trade, *Report,* p. 160.

53. U.K. Department of Energy, *Memorandum of Understanding* (London: HMSO, 1975), and *Code of Practice* (London: HMSO, 1975), p. 1.

54. *Banker,* May 1977, p. 89.

55. The resulting loss of catches (in 1976 prices) was assessed at between $90,000 and $828,000 in 1976, rising to between $126,000 and $1,080,000 by 1986. U.K. Department of Energy, *Development of Oil and Gas Resources,* pp. 20, 29–30, 37; interview #61 with an editor, *Fishing News,* London, 1978; and University of Aberdeen, Department of Political Economy and the Institute for the Study of Sparsely Populated Areas, *Loss of Access to Fishing Grounds Due to Oil and Gas Installations in the North Sea,* Research Report no. 1 (Aberdeen, 1978), pp. 126–28.

56. *Economist,* 21 January 1978.

57. Interview #67 with a high-level official, Scottish Fishermen's Federation, Scottish Division, Aberdeen, Scotland, 4 April 1978.

58. Interview #57 with a high-level official, FOOCG and Ministry of Agriculture, Fisheries, and Food, London, 21 March 1978.

59. Noreng, *The Oil Industry,* pp. 170–71; *Petroleum Economist,* January 1981, pp. 16–19.

60. *Petroleum Economist,* June 1982, pp. 232–33, and June 1983, p. 205.

61. *Economist,* 19 March 1982; and *Petroleum Economist,* April 1982, p. 133, and May 1983, p. 163.

62. *Economist,* 7 January 1984.

63. Britain's North Sea oil production was 1.57 million b/d by 1979. U.S. Department of Energy, Energy Information Administration, *Monthly Energy Review,* April 1981, p. 93; *World Business Weekly,* 7 July 1980, p. 16, and 27 October 1980, p. 16; and *Financial Times,* 12 October 1981.

64. *Economist,* 3 April 1982 and 2 April 1983.

65. *Economist,* 7 January 1984.

66. *Petroleum Economist,* November 1982, p. 449.

67. Ibid., and *Economist,* 7 January 1984.

68. *Petroleum Economist,* November 1982, p. 450, and May 1980, p. 214; *Economist,* 17 August 1985, 4 May 1986, and 31 May–6 June 1986.

69. *New York Times,* 3 December 1984.

70. *New York Times,* 8 and 31 December 1984; *Economist,* 7 January 1984; James Alt, "Oil and the Political Economy of Thatcherism," paper prepared for Conference on the Thatcher Government and British Political Economy (Harvard University Center for European Studies, 19–20 April 1985), p. 9; and *Economist,* 31 May–6 June 1986.

71. *Economist,* 22 September 1984.

72. Anderson G. Bartlett et al., *Pertamina: Indonesian National Oil* (Djakarta: Amerasian, 1972), pp. 198, 181, 325–29.

73. Interview #147 with former high-level official in Pertamina and OPEC, Jakarta, Indonesia, 19 September 1978.

74. J. R. V. Prescott, *The Political Geography of the Oceans* (Newton Abbott: David & Charles, 1975).

75. Interview #147.

76. Bartlett et al., *Pertamina,* pp. 382–84.

77. Kjell-Arne Ringbakk, "Multinational Planning and Strategy," Faculty Working Papers, Amos Tuck School of Business Administration (Philadelphia, 1975).

78. Bartlett et al., *Pertamina,* pp. 294–333.

79. Bartlett et al., *Pertamina,* p. 181.

80. Ibid., p. 236, and Alex Hunter, "The Indonesian Oil Industry," in Bruce Glassburner, ed., *The Economy of Indonesia* (Ithaca: Cornell University Press, 1971), p. 313.

81. Bartlett et al., *Pertamina,* p. 312.

82. Pertamina would invest in oil-related shipping, while government-run shipping would remain separate—sticking primarily to coastal or island trade.

83. Interview #147.

84. The IMF was primarily concerned that Pertamina was not following the rules of international finance and wanted to supervise the company's investments. Its major worry was over Pertamina's accounting procedures for the calculation of assets and, therefore, collateral to take out loans (interview #147). However, it was also concerned that company funds were being used for corrupt personal gains and political payoffs.

85. The U.S. government saw that oil shipments would be more secure if transported by the Majors rather than Pertamina, because Indonesia was a member of OPEC. Second, during 1972 the United States was importing about one-fifth of Indonesia's 300 million barrels of oil (*Asia Week,* 18 July 1978, p. 34), so Pertamina's oil politics could jeopardize a part of U.S. supplies. Third, the increasing power of Ibnu Sutowo, Pertamina's head, was a potential threat to the stability of the pro-U.S. Indonesian government and thus also to U.S. relations with other Southeast Asian countries.

86. Interview #143 with former high-level official, Ministry of Mines,

Jakarta, Indonesia, 12 September 1978; Bartlett et al., *Pertamina*, p. 325; Bruce Glassburner, "Indonesia's New Economic Policy and Its Sociopolitical Implications," in Karl Jackson and Lucien Pye, *Political Power and Communications in Indonesia* (Berkeley: University of California Press, 1978).

87. Interview #147, and interview #139 with consultant, Ministry of Finance, Jakarta, Indonesia, 8 September 1978.

88. Pertamina's production fell from 140,000 b/d in 1974, before the government's takeover of the company, to only 90,000 b/d in 1978. Interview #148 with high-level official in Inpex Indonesia Petroleum Ltd., Jakarta, Indonesia, 20 September 1978. Some observers contend that Pertamina, particularly since the 1975 reorganization of the company, has not exercised the managerial control it legally holds over foreign oil companies. Although I agree that the problem of actual versus theoretical state control is a serious one, I would argue that altered foreign decision making or foreign company preferences based on perceived managerial constraint are the kinds of evidence that show that the Indonesia state actually exercised control over foreign companies. For example, foreign oil companies have been constrained to subcontract Pertamina's (Pertamina Tongkang) supply boats when they would otherwise have contracted other foreign companies' vessels or used their own. Also, the Majors prefer to work under work contracts but have agreed to production-sharing arrangements to gain offshore leases from the Indonesian government.

89. *Petroleum Economist*, October 1981.

90. *Petroleum Economist*, December 1982, p. 514.

91. Inpex-Total had only been producing 5,000 b/d before 1975, but new onshore production brought the joint-venture company's production level up to 1.3 million b/d by March 1977. During the first six months of 1978 Inpex-Total production was averaging about 1.1 million b/d, placing the company far ahead of Caltex's 800,000 b/d during the same period. Indonesia's reported total crude oil production during the first half of 1978 was about 1.8 million b/d, a discrepancy of about 600,000 b/d from the sum of individual companies' production levels. Interview #148, 20 September 1978, with high-level official in a major foreign oil company working in Indonesia: "Production en Indonesie, 1976," an updated graph provided in the same interview. which shows the production level of all companies in Indonesia during 1973–78, including total production levels for the country.

92. Fishing accounted for only 2.1 percent of gross product in 1972. Lack of both capital and corporate structure deterred potential oil-related investments. Bartlett et al., *Pertamina*, p. 299; Hunter, "Indonesian Oil Industry," p. 311; Bartlett et al., *Pertamina*, pp. 382–83; and interview #130 with official from Fisheries Department, Ministry of Agriculture, Jakarta, Indonesia, 31 August 1978.

93. *Petroleum Economist*, February 1980, pp. 81–82.

94. *Petroleum Economist*, June 1982, p. 250.

95. *Bulletin of Indonesian Economic Studies*, July 1982, p. 3, and March

1982, p. 4; *Petroleum Economist,* June 1982, p. 249, and February 1983, p. 65; *OPEC Bulletin,* August 1982, p. 15; and *Quarterly Economic Review [QER] of Indonesia,* Annual Supplement, 1985, pp. 57–59.

96. *Petroleum Economist,* June 1982, p. 249.

97. *Petroleum Economist,* August 1984, p. 306.

98. Ibid.

99. *Petroleum Economist,* June 1982, p. 249, and *QER of Indonesia,* 1984, no. 1, p. 14.

100. *Petroleum Economist,* August 1984, pp. 305–6; *QER of Indonesia,* 1984, no. 3, p. 9.

101. Oil could make a greater contribution to gross national product in a shorter amount of time than either shipping or fishing. By the late 1960s the transportation and communication sector was only contributing 4 percent of GNP, while fishing was only about 3 percent. Besides, oil exports would improve Malaysia's position in the world economy. See David Lim, *Economic Growth and Development in West Malaysia, 1947–70* (Kuala Lumpur: Oxford University Press, 1973), and Prime Minister's Economic Planning Unit, *First Malaysian Plan, 1966–70, Mid-Term Review* (Kuala Lumpur, 1969), p. 14.

102. Malaysian Department of Information, *Malaysia 1971: Official Year Book* (Kuala Lumpur, 1972), p. 544; interview #106 with high-level official, Shell Malaysia, Kuala Lumpur, Malaysia, 7 August 1978; and *Quarterly Energy Review, Far East and Australia,* 1983, no. 2, p. 28.

103. Interview #108, 8 August 1978, with high-level official, Finance and Planning Division, Malaysian International Shipping Corporation (MISC), Kuala Lumpur, Malaysia and interview #125, 23 August 1978, with high-level official, Ministry of Trade, Kuala Lumpur, Malaysia.

104. Malaysian Ministry of Finance, *Economic Report, 1977/78* (Kuala Lumpur, 1978), p. 114, and interview #114, 10 August 1978, with an executive of Esso Malaysia, Kuala Lumpur, Malaysia.

105. Shipping posed little opposition to state oil enterprise because the industry was partly government-owned. Up to 1968 Malaysian shipping was virtually nonexistent. Most Malaysian trade was carried by foreign freight conferences, in particular the Far East Freight Conference, accounting for US$1.41 billion in exports and US$1.15 billion in imports by 1970. However, with conference rates increasing, the government joined with private sector interests in 1968 to form a national shipping company, MISC. In addition to trying to build a dry cargo fleet, MISC also made initial moves to expand into oil transportation. Japanese and Chinese capital gave the company the financial clout to make an entry into the Malaysian oil trade, but Shell refused to charter MISC's first tanker, so thwarting this attempt. Instead, Shell chartered Mobil tankers for all of its oil and, later, Esso relied mostly on the fleet of its parent company, Exxon. Malaysian Economic Planning Unit, *First Malaysia Plan,* p. 32; interview #108; interview #106.

106. Interview #128 with high-level official, MISC and Malaysian

Shipowners Association (MASA), Kuala Lumpur, Malaysia, 28 August 1978, and interview #108.

107. Interview #128.

108. Interview #128, and interview #125 with high-level official, Ministry of Trade, Kuala Lumpur, Malaysia, 23 August 1978.

109. Interview #128; interview #109 with official, Planning Division, Ministry of Transport, Kuala Lumpur, Malaysia, 9 August 1978; *QER of Malaysia*, Annual Supplement, 1983, p. 77.

110. Interview #114, and interview #108.

111. Interview #118 with a correspondent, *Asian Wall Street Journal*, Kuala Lumpur, Malaysia, 12 August 1978.

112. Interview #118, and interview #106 with a high-level executive, Shell Malaysia, Kuala Lumpur, Malaysia, 7 August 1978.

113. Interview #106. Similar to the Indonesian case, observers may claim that the Malaysian state's control was only tutelary vis-a-vis the foreign oil companies. However, that MISC LNG tankers will transport Malaysian gas exclusively when foreign oil companies would have retained more control through their own or other foreign tanker companies is evidence of the Malaysian state's ability to constrain foreign oil companies to act in a way other than they would act without such state control. See Ministry of Finance, *Economic Report*, p. 114, and interview #114.

114. Interview #115 with an executive, Esso Malaysia, Kuala Lumpur, Malaysia; 10 August 1978; interview #104 with a high-level official, Petronas, Kuala Lumpur, 2 August 1978, and interview #126 with an adviser, Prime Minister's Economic Planning Unit, Kuala Lumpur, Malaysia, 24 August 1978.

115. *QER of Malaysia*, Annual Supplement, 1984, p. 19.

116. Interview #100 with a high-level executive, Cities Service Mineral Company, Singapore, 28 July 1978; interview #107 with a high-level executive, Esso Malaysia, Kuala Lumpur, Malaysia, 8 August 1978; interview #114; and interview #106.

117. Interview #104 with a high-level official, Petronas, Kuala Lumpur, Malaysia, and interview #106.

118. Prime Minister's Economic Planning Unit, *Industrial Coordinative Act of the New Economic Policy*, adopted by Parliament 1971 (Kuala Lumpur, Malaysia).

119. Shell operations in East Malaysia were 45 miles from shore and covered 4,500 square miles off Sabah and 17,000 square miles off Sarawak. Esso's operations in West Malaysia were about 150 miles out, but covered 15,000 square miles. Interview #115 with field operations official, Esso Petroleum Malaysia, Kuala Lumpur, Malaysia, 10 August 1978; interview #106; interview #107 with exploration official, Esso Production Malaysia, Kuala Lumpur, Malaysia, 8 August 1978; interview #119 with high-level official, Fisheries Department, Kuala Trengganu, Malaysia, 14 August 1978; and *ICLARM Newsletter*, 1982, p. 20.

120. *Industrial Coordinative Act of the New Economic Policy,* and interview #119.

121. R. J. G. Wells, "Petroleum: Malaysia's New Engine of Growth?" *World Today,* July–August 1982, pp. 315–18; *Quarterly Energy Review,* p. 45; and *QER of Malaysia,* Annual Supplement, 1984, p. 113.

122. *Quarterly Energy Review,* no. 3, 1983, p. 22; *QER of Malaysia,* Annual Supplement, 1983, pp. 22, 79; and ibid., 1984, p. 13.

123. Wells, "Petroleum: Malaysia's New Engine," p. 317, and *QER of Malaysia,* 1983, no. 4, p. 15.

124. Business International Corporation, *Business Prospects in Malaysia: Coping with Change in a New Era,* Business International Research Report (Hong Kong, 1983), p. 85.

125. *QER of Malaysia,* 1983, no. 4, pp. 45, 77.

126. "Mahathir Tightens Grip on Cabinet and Party in Malaysian Reshuffle," *Business Asia,* 20 August 1984, p. 230; *QER of Malaysia, Brunei,* 1984, no. 4, p. 9, and 1985, no. 4, p. 9.

CHAPTER 4. *A Statist Interpretation of Oil Policy Choices*

1. See Thomas Neff, "The Changing World Oil Market," in David Deese and Joseph Nye, eds., *Energy and Security* (Cambridge, Mass.: Ballinger, 1981).

2. On the obsolescing bargain see Raymond Vernon, *Sovereignty at Bay: The Multinational Spread of U.S. Enterprises* (New York: Basic, 1971).

3. Norwegian Ministry of Industry, *Report No. 30 to the Norwegian Parliament (1973–74)* (unofficial government translation) pp. 76–84; M. M. Sibthrop, ed., *The North Sea: Challenge and Opportunity,* David Davies Memorial Institute of International Studies (London: Europa, 1975), pp. 255–58; and D. I. MacKay and G. A. MacKay, *The Political Economy of North Sea Oil* (London: Robertson, 1975), p. 25.

4. Interview #148 (graph provided) with high-level official in a major foreign oil company working in Indonesia, Jakarta, 20 September 1978. See List of Informants following Notes.

5. Interview #148 graph; Kjell-Arne Ringbakk, "Multinational Planning and Strategy," Faculty Working Papers, Amos Tuck School of Business Administration (Stanford, 1975), pp. 295–97.

6. Øystein Noreng, *The Oil Industry and Government Strategy in the North Sea* (London: Croom Helm, 1980), pp. 137–39.

7. Interview #147 with former high-level official in Pertamina and OPEC, Jakarta, Indonesia, 19 September 1978; Merrie Klapp, "Inter-Industry Conflict in the North Sea and South China Sea: A Comparative Analysis of Oil, Shipping and Fishing in Four Nations" (Ph.D. diss., University of California, Berkeley, 1980), p. 161; Anderson G. Bartlett et al., *Pertamina: Indonesian National Oil* (Djakarta: Amerasian, 1972), p. 325.

8. Klapp, "Inter-Industry Conflict," pp. 206–7.

9. Noreng, *The Oil Industry*, pp. 147–48.

10. Interview #141 with an executive, Arthur Young Associates (financial auditors for Pertamina), Jakarta, Indonesia, 12 September 1978; interview #142 with a high-level official, BAPPENAS (national economic planning agency), Jakarta, Indonesia, 12 September 1978; and interview #143 with former high-level official, Ministry of Mines, Jakarta, Indonesia, 12 September 1978.

11. Noreng, *The Oil Industry*, p. 150.

12. Interview #35 with high-level official, Economics and Legal Division, Petroleum Directorate, Stavanger, Norway, 7 February 1978.

13. Interview #143; interview #151 with high-level officials, Pertamina, Jakarta, Indonesia; and Klapp, "Inter-Industry Conflict," pp. 154–72.

14. Klapp, "Inter-Industry Conflict," p. 40.

15. *Economist*, 18 April 1981, pp. 70–71.

16. Interview #142.

17. Noreng, *The Oil Industry*, p. 147.

18. Klapp, "Inter-Industry Conflict," pp. 143–68.

19. Neff, "The Changing World Oil Market," pp. 26–27.

20. See Klapp, "Inter-Industry Conflict," sections on Norwegian and British fishing.

21. *Economist*, 15 October and 19 February 1983.

22. *Economist*, 19 February 1983.

23. *Economist*, 15 October 1983, p. 84.

24. Peter Cowhey, *The Problems of Plenty* (Berkeley: University of California Press, 1985), chap. 6.

25. *Petroleum Economist*, May 1980, p. 214, and November 1982, p. 450.

26. Cowhey, *Problems of Plenty*, p. 200.

27. Ibid., p. 186.

28. *Economist*, 18 April 1981, pp. 70–71.

29. *OPEC Bulletin* 13 (August 1982), pp. 14–22; *Petroleum Economist*, June 1982 and February 1983.

30. R. J. G. Wells, "Petroleum: Malaysia's New Engine of Growth?" *World Today*, July–August 1982, pp. 315–16.

31. *Economist*, 18 April 1981, and *Norwegian Bank Economic Bulletin*, 1982, no. 53, pp. 46–53.

32. *New York Times*, 3 December 1984.

33. *Petroleum Economist*, June 1982, and *Quarterly Economic Review [QER] of Indonesia*, 1984, no. 1, p. 14.

34. Raymond Vernon, "Linking Managers with Ministers: Dilemmas of the State-Owned Enterprise," *Journal of Policy Analysis and Management* 4, 1 (1984), p. 48.

35. *Economist*, 22 September 1984.

36. Bartlett et al., *Pertamina*, p. 393; *Petroleum Economist*, August 1984; and *QER of Indonesia*, 1984, no. 3, p. 9.

37. Cowhey, *Problems of Plenty*, p. 248.

38. Jeff Frieden, "Third World Indebted Industrialization: International Finance and State Capitalism in Mexico, Brazil, Algeria, and South Korea," *International Organization* 35 (Summer 1981).

39. *Economist*, 18 April 1981, pp. 70–71.

40. Frieden, "Third World Indebted Industrialization," p. 412; *QER of Indonesia*, Annual Supplement, 1985, p. 63, and 1980, p. 27.

41. Ibid., pp. 407, 414.

CHAPTER 5. *The Sovereign Entrepreneur Model: Implications for Other Countries*

1. See Karl Jackson, "Bureaucratic Policy: A Theoretical Framework for the Analysis of Power and Communications in Indonesia," in Jackson and Lucien Pye, eds., *Political Power and Communications in Indonesia* (Berkeley: University of California Press, 1978), p. 3.

2. On Mexico see especially Nora Hamilton, *The Limits of State Autonomy: Post-Revolutionary Mexico* (Princeton: Princeton University Press, 1982).

3. Here I draw from a point made by an anonymous reviewer of an early draft of this book.

4. Particularly useful on Iran and Saudi Arabia is Fariborz Ghadar, *The Evolution of OPEC Strategy* (Lexington: Heath, 1977).

5. Leslie Grayson, *National Oil Companies* (New York: Wiley, 1981), pp. 107–8.

6. Ibid., pp. 108–9, 134; Øystein Noreng, "State-Owned Oil Companies: Western Europe," in Raymond Vernon and Yair Aharoni, eds., *State-Owned Enterprise in the Western Economies* (New York: St. Martin's, 1980), pp. 133–42.

7. Dow Votaw, *The Six-Legged Dog* (Berkeley: University of California Press, 1964), p. 128.

8. Ibid., pp. 12, 20, 78; Grayson, *National Oil Companies*, p. 110.

9. On Italy see Grayson, *National Oil Companies*.

10. Helpful on developments in Japan is Richard Samuels, "State Enterprise, State Strength, and Energy Policy in Transwar Japan," MIT International Energy Studies Program, Energy Laboratory, MIT EL 83-010 WP (Cambridge, March 1982).

11. Anderson G. Bartlett et al., *Pertamina: Indonesian National Oil* (Djakarta: Amerasian, 1972), p. 234, and Samuels, "State Enterprise," pp. 44–45.

12. Samuels, "State Enterprise," pp. 48, 50–52.

13. On France see Grayson, *National Oil Companies*, and Harvey Feigenbaum, *The Politics of Public Enterprise* (Princeton: Princeton University Press, 1985).

14. Grayson, *National Oil Companies*, pp. 88–91; Samuels, "State Enterprise," p. 17; and Feigenbaum, *Politics of Public Enterprise*, pp. 50, 75.

15. Feigenbaum, *Politics of Public Enterprise*, pp. 14, 25; Grant McConnell, *Private Power and American Democracy* (New York: Knopf, 1966).

16. Ghadar, *Evolution of OPEC Strategy*, p. 59.

17. Grayson, *National Oil Companies*, p. 9, and Ghadar, *Evolution of OPEC Strategy*, p. 60.

18. Grayson, *National Oil Companies*, p. 178.

19. For the U.S. case, I draw heavily on Irvine Anderson, *Aramco, the United States and Saudi Arabia* (Princeton: Princeton University Press, 1981); Robert Keohane, "State Power and Industry Influence: American Foreign Policy in the 1940s," *International Organization* 36 (Winter 1982); and Aaron Miller, *Search for Security: Saudi Arabian Oil and American Foreign Policy, 1939–1949* (Chapel Hill: University of North Carolina Press, 1980).

CHAPTER 6. *Contrasting Perspectives on the Role of the State*

1. Peter Cowhey, *The Problems of Plenty* (Berkeley: University of California Press, 1985), Introduction.

2. Cf. Stephen D. Krasner, *Structural Conflict* (Berkeley: University of California Press, 1985), pp. 5–6, 7–10.

3. Charles Lipson, *Standing Guard* (Berkeley: University of California Press, 1985), pp. 140–89.

4. Theda Skocpol, "Bringing the State Back In: Strategies of Analysis in Current Research," in Peter Evans, Dietrich Rueschemeyer, and Skocpol, *Bringing the State Back In* (Cambridge: Cambridge University Press, 1985), p. 21.

5. Peter Katzenstein, "Small Nations in an Open International Economy: The Converging Balance of State and Society in Switzerland and Austria," in Evans et al., *Bringing the State Back In*, p. 239.

6. Peter Katzenstein, *Small States in World Markets* (Ithaca: Cornell University Press, 1985), pp. 24–58.

7. Dietrich Rueschemeyer and Peter Evans, "The State and Economic Transformation," in Evans, Rueschemeyer, and Skocpol, *Bringing the State Back In*, pp. 59–60.

8. Stephen D. Krasner, *Defending the National Interest: Raw Materials Investments and U.S. Foreign Policy* (Princeton: Princeton University Press, 1978), p. 11.

9. Skocpol, "Bringing the State Back In," p. 17.

10. Rueschemeyer and Evans, "The State and Economic Transformation," p. 59.

11. Theodore Moran, *Multinational Corporations and the Politics of Dependence* (Princeton: Princeton University Press, 1974), p. 155.

12. Peter Evans, *Dependent Development* (Princeton: Princeton University Press, 1979), pp. 9–11, 214, 273–85, and Fernando Cardoso and Enzo Faletto, *Dependency and Development in Latin America* (Berkeley: University of California Press, 1979), pp. 153, 165.

13. Katzenstein, "Small Nations," pp. 231–45; *Corporatism and Change* (Ithaca: Cornell University Press, 1984).

14. Katzenstein, *Small States*, pp. 24–58.

15. Hugh Heclo, *Modern Social Politics in Britain and Sweden* (New Haven: Yale University Press, 1974), pp. 303–6.

16. Cf. Martin Shefter, "Party and Patronage: Germany, England, and Italy," *Politics and Society* 7 (1977), and his "Party, Bureaucracy, and Political Change in the United States," in Louis Maisel and Joseph Cooper, eds., *The Development of Political Parties: Patterns of Evolution and Decay*, vol. 4 of *Sage Electoral Yearbook* (Beverly Hills, Calif.: Sage, 1979).

17. Alfred Stepan, *The State and Society: Peru in Comparative Perspective* (Princeton: Princeton University Press, 1978), chaps. 3 and 4; Karl Jackson and Lucian Pye, eds., *Political Power and Communications in Indonesia* (Berkeley: University of California Press, 1978).

List of Informants

One hundred fifty-six interviews were conducted primarily in Norway, Britain, Indonesia, and Malaysia between January and November 1978. These interviews support the explanation of events between 1965 and 1978 advanced in this book. Earlier and later periods were researched primarily using documents and other secondary materials. I have referred to specific interviews within the text only when a point was surprising or controversial, or the informant was the unique source of the information. For more complete references to informants in relation to my basic research, see Merrie Klapp, "Inter-Industry Conflict: A Comparative Analysis of Oil, Shipping and Fishing in Four Nations" (Ph.D. diss., University of California, Berkeley, 1980).

1–2. Initial interviews for a U.S. case study.

3. Foreign affairs specialist, International Fisheries Analysis Branch, National Oceanic and Atmospheric Administration, Washington, D.C.

4. High-level executive, Exploration Affairs, American Petroleum Institute, Washington, D.C.

5. Executive, Refining Department, American Petroleum Institute, Washington, D.C.

6. Fisheries attaché, Embassy of Norway, Washington, D.C.

7. Official, Office of Ocean Affairs, Department of State, Washington, D.C.

8. Professor, Center for Social Studies, University of Bergen, Bergen, Norway.

9. High-level official, Legal Division, Directorate of Fisheries, Bergen, Norway.

10. High-level official, Institute of Marine Research, Directorate of Fisheries, Bergen, Norway.

11. Researcher, Sociology Institute, University of Bergen, Bergen, Norway.

12. Consultant, Directorate of Fisheries, Bergen, Norway.

13. Consultant, Directorate of Fisheries, Bergen, Norway.

14. Officer, Office of Information, Directorate of Fisheries, Bergen, Norway.

15. Professor, Industrial Economics Institute, School of Economics, Bergen, Norway.

16. High-level official, Fishermen's Union, Bergen, Norway.

17. Consultant, Economics Division, Directorate of Fisheries, Bergen, Norway.

18. Executive, Westfal-Larsen & Co. A/S Shipowners, Bergen, Norway.

19. Consultant, Economics Division, Directorate of Fisheries, Bergen, Norway.

20. Researcher, Shipping Economics Institute, Bergen, Norway.

21. High-level official, Norges Sildeslag (fish processors for North Sea stocks), Bergen, Norway.

22. Executive, Norsildmel (fishmeal and fish oil sales organization), Bergen, Norway.

23. High-level consultant, Directorate of Fisheries, Bergen, Norway.

24. High-level official, Rogaland District Fishing Organization, Karmøy, Norway.

25. High-level officer, Sør Norges Tralerlag (trawlers' association for North Sea), Karmøy, Norway.

26. Researcher, Stavanger, Norway.

27. Senior Engineer, Petroleum Directorate, Stavanger, Norway.

28. Senior legal counselor, Petroleum Directorate, Stavanger, Norway.

29. Geologist, Public Affairs and Information Department, Statoil, Stavanger, Norway.

30. Executive, Peter Smedvig Shipping Co., Stavanger, Norway.

31. Editor, Noroil Publishing House, Stavanger, Norway.

32. Executive, Joint Ventures Division, Elf Aquitaine Norge A/S, Stavanger, Norway.

33. High-level officer, Rogaland Fiskarlag (fishing union for Rogaland), Stavanger, Norway.

34. Consultant, Planning Commission, Stavanger, Norway.

35. High-level official, Economics and Legal Division, Petroleum Directorate, Stavanger, Norway.

36. Lawyer, Norwegian Offshore Association, Oslo, Norway.

37. Executive, De Norske Sildolle- og Sildemelfabrikkers Landsforening (fish oil and fishmeal), Oslo, Norway.

38. Consultant, Ministry of Commerce and Shipping, Oslo, Norway.

39. High-level official, Ministry of Commerce and Shipping, Oslo, Norway.

40. High-level official, Norwegian Shipowners' Association, Oslo, Norway.

41. Executive, R. S. Platou A/S (shipping brokerage firm), Oslo, Norway.

42. High-level official, Ministry of Industry, Oslo, Norway.

43. Professor, Law School, University of Oslo, Oslo, Norway.

44. High-level executive, Stad Seaforth Shipping Co., Oslo, Norway.

45. Consultants, Research Department and Supply Division of Offshore Department, Fearnley & Egers Chartering Co., Oslo, Norway.

46. Consultant, Offshore Department, Fearnley & Egers Chartering Co., Oslo, Norway.

47. High-level official, Petroleum Legislation, Petroleum Directorate, Oslo, Norway.

48. Official, Norwegian Seamen's Union, Oslo, Norway.

49. Counselor, Ministry of Oil and Energy, Oslo, Norway.

50. Official, Economics Division, Ministry of Fisheries, Oslo, Norway.

51. Official, Quotas Division, Ministry of Fisheries, Oslo, Norway.

52. High-level official, International Council for Exploration of the Seas, Charlottenlund, Denmark.

53. Official, conservation issues, American Embassy, London, England.

54. Official, Quotas Division, Ministry of Agriculture, Fisheries and Food, London, England.

55. Official, Statistics Division, Ministry of Agriculture, Fisheries and Food, London, England.

56. Official, Oil and Fishing Problems, Ministry of Agriculture, Fisheries and Food, London, England.

57. High-level official, Fishing and Offshore Oil Consultative Group and Ministry of Agriculture, Fisheries and Food, London, England.

58. Professor, Political Science, City University, London, England.

59. High-level official, Department of Energy, London, England.

60. High-level official, Department of Trade and Industry, London, England.

61. High-level editor, *Fishing News*, London, England.

62. Official, Statistics Division, Department of Trade and Industry, London, England.

63. High-level executive, General Council of British Shipping, London, England.

64. Official, National Union of Seamen, London, England.

65. High-level official, National Union of Seamen, London, England.

66. Professor, Department of Political Economy, University of Aberdeen, Aberdeen, Scotland.

67. High-level official, Scottish Fishermen's Federation, Aberdeen, Scotland.

68. Official, Department of Fisheries for Scotland, Aberdeen, Scotland.

69. Official, Scottish Trawlers' Federation, Aberdeen, Scotland.

70. Executive, Offshore Marine, Aberdeen, Scotland.

71. Executive, Seaforth Marine, Aberdeen, Scotland.

72. Professor, Department of Sociology, University of Aberdeen, Aberdeen, Scotland.

73. High-level official, Transport and General Workers Union, Aberdeen, Scotland.

74. High-level official, Scottish Fishermens' Federation, Fraserburgh, Scotland.

75. High-level official, Whitefish Association, Peterhead, Scotland.

76. High-level official, British Fishermens' Federation, Scottish Division, Aberdeen, Scotland.

77. High-level executive, Hall Russell (shipbuilders), Aberdeen, Scotland.

78. Executive, Shell U.K., Aberdeen, Scotland.

79. High-level official, Fish Marketing Association, Aberdeen, Scotland.

80. Executive, ODECO U.K. (offshore drilling and exploration company), Aberdeen, Scotland.

81. High-level executive, John Wood Group (fishing, fishing boat repair, and oil service company), Aberdeen, Scotland.

82. High-level official, Offshore Supplies Office, Aberdeen, Scotland.

83. Official, Department of Agriculture and Fisheries for Scotland, Edinburgh, Scotland.

84. Official, Whitefish Authority, Edinburgh, Scotland.

85. Official, Department of Agriculture and Fisheries for Scotland, Edinburgh, Scotland.

86. Official, Statistics, Department of Agriculture and Fisheries, Edinburgh, Scotland.

87. High-level official, Department of Agriculture and Fisheries for Scotland, Edinburgh, Scotland.

88. High-level official, Fishing and Offshore Oil Consultative Group, Department of Agriculture and Fisheries, Edinburgh, Scotland.

89. High-level official, Offshore Supplies Office, Glasgow, Scotland.

90. Professor, University of Reading, Reading, England.

91. High-level official, U.K. Offshore Operators Association, London, England.

92. High-level executive, P&O Shipping Co., London, England.

93. Official, Inter-governmental Maritime Consultative Organization, London, England.

94. Official, Inter-governmental Maritime Consultative Organization, London, England.

95. Official, Inter-governmental Maritime Consultative Organization, London, England.

96. Official, U.N. Food and Agriculture Organization, Rome, Italy.

97. Official, U.N. Food and Agriculture Organization, Rome, Italy.

98. High-level official, Shipping Division, Economic and Social Commission for Asia and the Pacific, Bangkok, Thailand.

99. High-level official, Committee for the Coordination of Joint Prospecting for Mineral Resources in Asian Offshore Areas, Bangkok, Thailand.

100. High-level executive, Cities Service Mineral Co., Singapore.

101. Fisheries Economist, Department of Economics and Administration, University of Malaya, Kuala Lumpur, Malaysia.

102. Official, Fisheries Division, Ministry of Agriculture, Kuala Lumpur, Malaysia.

103. Official, Fisheries Division, Ministry of Agriculture, Kuala Lumpur, Malaysia.

104. High-level official, Petronas, Kuala Lumpur, Malaysia.

105. High-level official, Marketing Division, Fisheries Development Authority, Kuala Lumpur, Malaysia.

106. High-level executive, Shell Malaysia, Kuala Lumpur, Malaysia.

107. High-level executive, Esso Malaysia, Kuala Lumpur, Malaysia.

108. High-level official, Finance and Planning Division, Malaysian International Shipping Corporation, Kuala Lumpur, Malaysia.

109. Official, Planning Division, Ministry of Transport, Kuala Lumpur, Malaysia.

110. Official, Planning Division, Ministry of Transport, Kuala Lumpur, Malaysia.

111. High-level official, Shipping Division, Ministry of Transport, Kuala Lumpur, Malaysia.

112. Official, Intergovernmental Maritime Consultative Organization representative at Southeast Asian Agency for Regional Transport and Communications Development, Kuala Lumpur, Malaysia.

113. High-level official, International Law Division, Attorney General's Chambers, Kuala Lumpur, Malaysia.

114. Executive, Esso Malaysia, Kuala Lumpur, Malaysia.

115. Executive, Esso Malaysia, Kuala Lumpur, Malaysia.

116. Official, National Secretariat for Association of South-East Asian Nations, Ministry of Foreign Affairs, Kuala Lumpur, Malaysia.

117. Official, Association of South-East Asian Nations Secretariat, Ministry of Foreign Affairs, Kuala Lumpur, Malaysia.

118. Correspondent, *Asian Wall Street Journal*, Kuala Lumpur, Malaysia.

119. High-level official, Fisheries Department, Kuala Trengganu, Malaysia.

120. Official, Marine Department, Kuala Trengganu, Malaysia.

121. Consultant, Shipping and Ports Division, Economic and Social Commission for Asia and the Pacific, Bangkok, Thailand.

122. High-level official, Information Division, Economic and Social Commission for Asia and the Pacific, Bangkok, Thailand.

123. High-level official, Transnational Corporate Study Group, Economic and Social Commission for Asia and the Pacific, Bangkok, Thailand.

124. High-level official, Committee for the Coordination of Joint Prospecting for Mineral Resources in Asian Offshore Areas, Bangkok, Thailand.

125. High-level official, Ministry of Trade, Kuala Lumpur, Malaysia.

126. Adviser, Prime Minister's Economic Planning Unit, Kuala Lumpur, Malaysia.

127. High-level official, Finance Division, Treasury, Kuala Lumpur, Malaysia.

128. High-level official, Malaysian International Shipping Corporation and Malaysian Shipowners Association, Kuala Lumpur, Malaysia.

129. High-level official, Malaysian Trade Union Conference, Kuala Lumpur, Malaysia.

130. Consultant, Fisheries Department, Ministry of Agriculture, Jakarta, Indonesia.

131. Consultant, Directorate General of Fisheries, Ministry of Agriculture, Jakarta, Indonesia.

132. High-level executive, Atlantic Richfield Indonesia, Jakarta, Indonesia.

133. High-level official, Directorate of Enterprise Development, Directorate General of Fisheries, Jakarta, Indonesia.

134. High-level official, Exploration Department, Pertamina, Jakarta, Indonesia.

135. High-level official, Directorate General of Sea Communication, Jakarta, Indonesia.

136. Official, Government Enterprises Division, Directorate General of Fisheries, Jakarta, Indonesia.

137. Official, Enterprise Development, Directorate General of Fisheries, Jakarta, Indonesia.

138. Officer, Economic Section, U.S. Embassy, Jakarta, Indonesia.

139. High-level consultant, Ministry of Finance, Jakarta, Indonesia.

140. United Nations Development Program representative, Department of Fisheries, Jakarta, Indonesia.

141. Executive, Arthur Young Associates (financial auditors for Pertamina), Jakarta, Indonesia.

142. High-level official, BAPPENAS (national economic development planning agency), Jakarta, Indonesia.

143. Ex-high-official, Ministry of Mines, Jakarta, Indonesia.

144. High-level official, Pertamina Directorate of Shipping and Telecommunications, Jakarta, Indonesia.

145. High-level official, Bureau of Science and Technology, Association of South-East Asian Nations Secretariat, Jakarta, Indonesia.

146. Correspondent, *Asian Wall Street Journal*, Jakarta, Indonesia.

147. Ex-high-level official, Pertamina and Organization of Petroleum-Exporting Countries, Jakarta, Indonesia.

148. High-level executive, INPEX (Indonesia Petroleum Co.), Jakarta, Indonesia.

149. Official, Division for Coordination of Foreign Contractors, Pertamina, Jakarta, Indonesia.

150. High-level official, Research and Development, Directorate General of Sea Communications, Jakarta, Indonesia.

151. High-level officials, Pertamina, Jakarta, Indonesia.

152. Ex-high-level official, Directorate General of Sea Communications, Jakarta, Indonesia.

153. High-level official, Director of Resources Management, Directorate General of Fisheries, Jakarta, Indonesia.

154. High-level official, Asian Development Bank, Manila, Philippines.

155. High-level official, Iclarm, Manila, Philippines.

156. Professor, International Finance, Department of Economics, University of California, Santa Barbara, California.

Index

Library of Congress Cataloging-in-Publication Data

Klapp, Merrie Gilbert, 1950-
   The sovereign entrepreneur.

   (Cornell studies in political economy)
   Includes index.
   1. Petroleum industry and trade—Government policy—Case studies.
2. Petroleum industry and trade—Government ownership—Case studies.   I. Title.   II. Series.
HD9560.6.K42   1987   338.2'7282'091724   86-19909
ISBN 0-8014-1997-2 (alk. paper)